PHYSIOLOGY OF NEMATODES

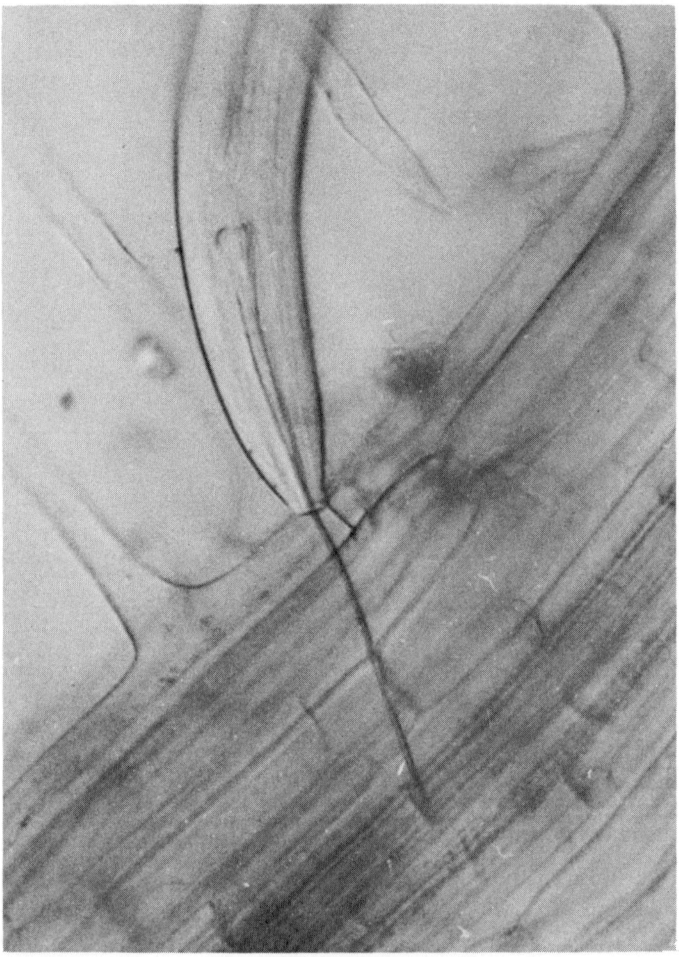

Frontispiece Plant-parasitic nematode (*Xiphinema*) feeding by means of its stylet upon the contents of the vascular tissue in a plant root. (Courtesy of D. L. Trudgill.)

PHYSIOLOGY OF NEMATODES

D. L. LEE and H. J. ATKINSON

Second Edition

New York Columbia University Press 1977

ERRATA

Page 6, line 16, and page 9, lines 7 and 18: *for* figure 1.5 *read* figure 6.2
Page 8, lines 6, 12 and 16; page 9, lines 2 and 4: *for* figure 1.6 *read* figure 1.5
Page 9, lines 18 and 22: *for* figure 6.2 *read* figure 6.3
Page 9, line 26: *for* figure 6.3 *read* figure 6.4
Page 9, line 27: *for* figure 6.4 *read* figure 6.5
Page 9, line 34: *for* figure 6.5 *read* figure 6.6
Page 30, Page 30, line 4 from bottom of page, and page 39, line 7 from bottom of page: *for* figure 1.5 *read* frontispiece

D. L. Lee is Professor of Agricultural Zoology and H. J. Atkinson is a Lecturer in Agricultural Zoology in the Department of Pure and Applied Zoology, University of Leeds.

First published in Great Britain in 1965 by Oliver and Boyd Ltd

Second Edition published in 1976 in Great Britain by
The Macmillan Press Ltd and in 1977 in the United States of America by
Columbia University Press

Printed in Great Britain

Library of Congress Cataloging in Publication Data

Lee, Donald Lewis.
Physiology of nematodes.

Bibliography: p.
Includes index.
1. Nematoda – Physiology. I. Atkinson, H. J., joint author. II. Title.
QL391.N4L37 1977 595'.182'041 77–1232
ISBN 0–231–04358–9

Contents

Preface ix

1. Introduction 1
1.1. General Biology 1
1.2. The Organisation of the Nematode Body 3
1.2.1. Cuticle 3
1.2.2. Hypodermis 4
1.2.3. Muscles 5
1.2.4. Nervous system 6
1.2.5. Pseudocoelom 7
1.2.6. Pseudocoelomocytes 7
1.2.7. Alimentary system 7
1.2.8. Excretory system 8
1.2.9. Reproductive system 9
1.2.10. Hypodermal, caudal and rectal glands 10

2. Cuticle, Moulting and Growth 11
2.1. The Cuticle 11
2.1.1. Structure 11
2.1.2. Composition 14
2.2. Moulting and Ecdysis 15
2.2.1. Introduction 15
2.2.2. The stimulus 15
2.2.3. Formation of the new cuticle and ecdysis 16
2.2.4. Neurosecretory control of moulting and ecdysis 21
2.2.5. Ecdysis of ensheathed juveniles 22
2.3. Growth 25

3. Feeding and Digestive Physiology 28
3.1. The Alimentary Canal 28
3.1.1. Stomodaeum 28
3.1.2. The functioning of the pharynx as a pump 34
3.1.3. Intestine 36
3.1.4. Proctodaeum 38
3.2. Feeding 38
3.2.1. Microbivorous and saprophagous 38
3.2.2. Phytophagous and mycophagous 39
3.2.3. Carnivorous 43

3.2.4. Entomophagous 44
3.2.5. Parasites of vertebrates 45
3.3. Digestive Enzymes 48
3.3.1. Enzymes in the pharynx 48
3.3.2. Enzymes in the intestine 50
3.3.3. Enzymes in the sub-ventral (excretory) glands 52
3.4. Secretion in the Intestine 53
3.5. Uptake of Nutrients 53
3.6. Nutrient Requirements and Growth *In Vitro* 55
3.7. Defecation 56
3.8. Anti-enzymes and Anti-coagulants 56

4. Metabolism 58
4.1. Carbohydrate Metabolism 58
4.1.1. Distribution of carbohydrates 58
4.1.2. Intermediary metabolism 60
4.1.3. Glycolysis: the Embden–Meyerhof pathway 62
4.1.4. Pasteur effect 63
4.1.5. Production of lactic acid and the oxygen debt 63
4.1.6. Oxidative decarboxylation: the tricarboxylic acid cycle 64
4.1.7. The TCA cycle in nematodes 64
4.1.8. Alternative pathways of carbohydrate metabolism in nematodes 65
4.1.9. The electron transport system: oxidative phosphorylation 70
4.1.10. The electron transport system in nematodes 70
4.1.11. Electron transport in *Ascaris* 72
4.1.12. P/O ratio 73
4.1.13. The pentose–phosphate pathway 74
4.1.14. Glycogenesis and gluconeogenesis 75
4.1.15. Nucleotides and nucleosides 75
4.1.16. Production of organic acids 76
4.2. Lipid Metabolism 76
4.2.1. Distribution of lipids 76
4.2.2. Lipid metabolism 79
4.3. Protein Metabolism 80
4.3.1. Distribution of proteins 80
4.3.2. Protein metabolism 83
4.4. Respiratory Physiology 85
4.4.1. Diffusion of oxygen 85
4.4.2. Factors which influence oxygen demand 86
4.4.3. Availability of oxygen 88
4.4.4. Respiration in low-oxygen regimes 89
4.4.5. Haemoglobin 91

4.5. Pharmacology 93
4.5.1. Glucose uptake and glycolysis 95
4.5.2. TCA cycle, succinate formation and phosphorylation 95

5. Osmotic and Ionic Regulation: Excretion 97
5.1. Osmotic Regulation 97
5.1.1. Introduction 97
5.1.2. Volume regulation in nematodes 97
5.1.3. Permeability to water 103
5.1.4. Desiccation survival 103
5.2. Ionic Regulation 104
5.2.1. Introduction 104
5.2.2. Ionic regulation in nematodes 105
5.2.3. Permeability to solutes 107
5.3. Excretion 109
5.3.1. Introduction 109
5.3.2. Products of nitrogen catabolism in nematodes 109
5.3.3. Other excretory products 112
5.3.4. Excretion of nitrogenous waste 112
5.3.5. Functions of the excretory system 113

6. Reproductive Physiology and Hatching 116
6.1. Reproductive Physiology 116
6.1.1. The male system and spermatogenesis 116
6.1.2. Sex attraction 119
6.1.3. Copulation 120
6.1.4. The female system and oogenesis 120
6.1.5. Fertilisation 121
6.1.6. Structure and formation of the egg shell 122
6.1.7. Egg laying 125
6.2. Hatching 125
6.2.1. Emergence of juveniles from eggs 125
6.2.2. General environmental effects on hatching 127
6.2.3. Specific hatching stimuli: temperature 128
6.2.4. Specific hatching stimuli: chemical 129

7. Neuro-muscular Physiology 133
7.1. Muscles 133
7.2. The Nervous System 136
7.2.1. Structure 136
7.2.2. The nerve–muscle junction 138
7.2.3. Transmission along nerves 138
7.2.4. Neuro-muscular and synaptic transmission 138
7.3. Pharmacology 142

8. Locomotion 145
8.1. The Hydrostatic Skeleton 145
8.2. Sinusoidal Movement 147
8.2.1. Swimming 147
8.2.2. Movement in water films (crawling) 149
8.2.3. Movement in soils 152
8.2.4. Movement within animals 155
8.2.5. Movement within the egg 155
8.3. Other Forms of Movement 156
8.4. Dispersal Mechanisms 158

9. Sense Organs and Behaviour 161
9.1. Sense Organs 161
9.1.1. Introduction 161
9.1.2. Mechanoreceptors 161
9.1.3. Chemoreceptors 164
9.1.4. Photoreceptors 167
9.2. Behaviour 168
9.2.1. Responses to chemicals 169
9.2.2. Response to mechanical stimulation 172
9.2.3. Reactions to heat 172
9.2.4. Responses to light 173
9.2.5. Responses to an electrical field 174
9.2.6. Responses to gravity 175
9.2.7. Response to a moisture gradient 175
9.3. Whole Patterns of Behaviour 175
9.3.1. Location and penetration of host plants 176
9.3.2. Location and penetration of host animals 177

Appendix 179
References 187
Index 200

Preface

At one time nematodes were studied almost exclusively by plant pathologists and by parasitologists, but they are now also studied by biochemists, biophysicists, geneticists, immunologists, molecular biologists, pharmacologists and neurophysiologists, who see the nematode as an exciting and challenging animal suitable for many fields of fundamental and applied research.

During the ten years that have elapsed since the appearance of the first edition of this book, a large number of research papers and reviews and several books, which describe various aspects of the biochemistry, physiology, functional morphology and ecology of free-living and parasitic nematodes, have been published. This book is intended to be an introduction to the physiology, biochemistry and structure of nematodes for undergraduate students and their teachers, and is not intended as an exhaustive review of the subject, although it is hoped that research workers in the field may also find it useful.

The information in this book is based on many publications. Unfortunately, it has not been possible to include references to all of them; consequently many references are to books, reviews and recent papers and these will introduce the reader to the earlier, more detailed, literature. For this reason, reference to an author in the text does not imply that this author was the first to have made the observation in question.

We are indebted to Professor A. O. Anya, Dr K. S. Cheah, Dr D. J. McLaren, Mr D. R. Newall, Dr D. J. Wright and Dr K. A. Wright, who allowed us to read some of their work prior to publication, and to Mrs Marjorie Miles who typed the manuscript.

We are also indebted to the following for permission to reproduce illustrations and tables: Academic Press Inc. for figure 5.4 from *Experimental Parasitology*, figure 2.3 from *The Structure of Nematodes* by A. F. Bird, figure 2.6 from *The nature of Parasitism* by W. F. Rogers, figure 4.3 from *Comparative Biochemistry of Parasites* edited by H. Van den Bossche, figure 7.5 from *Chemical Zoology* vol. 3 edited by M. Florkin and B. T. Scheer, figures 5.1, 5.2 and 5.6 from *The Organization of Nematodes* edited by N. A. Croll, table 4.2 from *Advances in Parasitology* and table 2.1 from *Archives of Biochemistry and Biophysics;* Cambridge University Press for figures 1.3, 2.2, 2.7, 3.6, 3.7, 5.3 and 8.6 from *Parasitology*; Association of Applied Biologists for figures 6.9, 8.5 and 9.7 from *Annals of Applied Biology*; The Society for Experimental Biology for figure 5.5 from *Symposia of the Society for Experimental Biology* vol. 23; The Company of Biologists Ltd, for figure 2.9 from *The Quarterly Journal of Microscopical Science* and figures 4.10 and 8.3 from *The Journal of Experimental Biology*; Dr M. B. Chitwood for figure 1.5 and part of figure 3.3 from *An Introduction to Nematology* by B. G. and M. B. Chitwood; Hutchinson & Co. for figures 3.6 and 7.4*B* from *Nematodes* by H. D. Crofton; E. J. Brill Ltd, for figures 6.8, 9.3 and 9.8 from *Nematologica*; The University of North Carolina Press for figures 4.6 and 4.9 from *Nematology. Fundamentals and Recent Advances with Emphasis on Plant Parasitic and Soil Forms* edited by J. N. Sasser and W. R. Jenkins; the Controller of Her Majesty's Stationery Office for part of figures 3.3 and 9.1 from *Plant Nematology* edited by J. F. Southey; The American Society of Tropical Medicine and Hygiene for figure 7.6 from *The American Journal of Tropical Medicine and Hygiene*; The Washington Academy of Sciences for figure 9.6 from *Journal of the Washington Academy of Sciences*; Helminthological Society of Washington for figure 7.1 from *Proceedings of the Helminthological Society of Washington*; Oliver & Boyd Ltd, for figures 2.4 and 2.5 from *Tissue and Cell*; The Rockefeller University Press for figures 7.2 and 7.3 from the *Journal of Cell Biology*; The Wistar Institute Press for figures 9.2, 9.4 and 9.5 from *The Journal of Comparative Neurology* and for figure 1.6 from *Journal of Morphology*; The American Society of Parasitology for part of figure 4.9 from *The Journal of Parasitology*; R. B. Clark for part of figure 8.1 from *Dynamics of Metazoan Evolution*; Annual Reviews Inc. for figures 3.8, 8.2, 8.4 and 8.9 from *Annual Reviews of Phytopathology*; H. R. Wallace and Edward Arnold (Publishers) Ltd., for figure 8.7 from *The Biology of Plant Parasitic Nematodes*; The American Society of Zoologists for figure 7.4 from *American Zoologist.* The U.S. National Academy of Sciences for table 9.2 from *Proceedings of the National Academy of Sciences, U.S.A.*; W. B. Saunders Co. and C. L. Prosser for table 9.1 from *Comparative Animal Physiology*; Birkhäuser Verlag for figure 7.5 from *Experimentia.*

Acknowledgements are made to authors by mentioning their names in the legend of the figures.

December 1975

D. L. Lee
H. J. Atkinson

1 Introduction

1.1 General Biology

There is a general belief among biologists that nematodes are mainly parasitic, however large numbers of free-living nematodes inhabit marine- or fresh-water-mud and the soil and several million can be found per square metre in the top few centimetres of sub-littoral mud. These free-living nematodes feed upon bacteria, fungi, decaying materials and microscopic animals. They are seldom seen, however, as most of them are less than a millimetre in length. Nematodes are widely distributed and are found in almost every type of environment.

> They occur in arid deserts and at the bottom of lakes and rivers, in the waters of hot springs and in the polar seas where the temperature is constantly below the freezing point of fresh water. They were thawed out alive from Antarctic ice in the far south by members of the Shackleton Expedition. They occur at enormous depths in Alpine lakes and in the ocean. As parasites of fishes they traverse the seas; as parasites of birds they float across continents and over high mountain ranges.[1]

Some will survive for many months in liquid nitrogen; others live in unusual habitats such as vinegar, bill posters' paste or drip mats that have been soaked in beer. A number of species are cryptobiotic and can survive for months or years in an almost desiccated state until water becomes available, when they revive. Nematodes also feed on living plants, either as ecto- or endoparasites, and they are one of the important groups of invertebrate parasites of animals and man.

The typical nematode is spindle-shaped, unsegmented and bilaterally symmetrical (figures 1.1 and 1.2). The adoption of a parasitic life has, in most cases, done little to alter the general shape of nematodes and

> . . . the elementary student may be forgiven at times for thinking that there is only one nematode, but that the model comes in different sizes and with a great variety of life histories.[2]

Respiratory and circulatory organs are absent and the excretory organs are unlike those of other invertebrates. Typical flagella have been found in the intestine of *Eudorylaimus*[3] but this is a rare exception as motile cilia and flagella are not normally present in nematodes. Modified flagella are present in some sense organs. The body wall is composed of a collagenous cuticle, a cellular or syncytial hypodermis, and a layer of longitudinal muscle. The alimentary tract consists of a terminal mouth, buccal cavity, pharynx (more usually called the oesophagus), intestine, rectum (cloaca in the male), and sub-terminal anus. The body cavity, in which lie the gonads, is a pseudocoelom (figures 1.1 and 1.2).

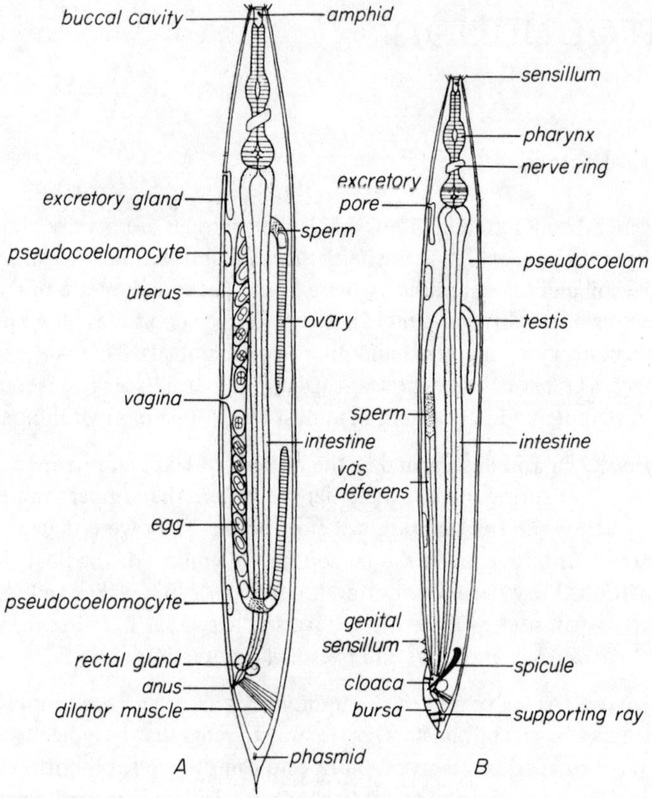

Figure 1.1 General morphology of a nematode (hypothetical). (*A*) female; (*B*) male. Lateral view.

The sexes are almost always separate. Parthogenetic females occur in some species and in others hermaphrodites are found, the gonads first producing sperm and later producing eggs, but these are not typical. The males are usually smaller than the females and possess copulatory aids such as genital sensilla and spicules. The sperm are amoeboid and different groups of nematodes have sperm of a characteristic shape and structure (see figure 6.3).

Nematodes undergo four moults from egg to adult worm. The juvenile (often referred to as a larva in the literature) is similar to the adult nematode but differs from it in size, sometimes in mouth parts, in the lack of gonads and the absence of copulatory structures. Some nematodes moult once or twice within the egg before hatching; others retain the cuticle of the previous stage at a certain stage in the life cycle. This acts as a protective covering against adverse conditions, as in infective juveniles of various animal-parasitic species of trichostrongyles and in 'dauer' juveniles of some free-living species that are transported from place to place on the outside of insects. After the fourth moult the nematode emerges as the

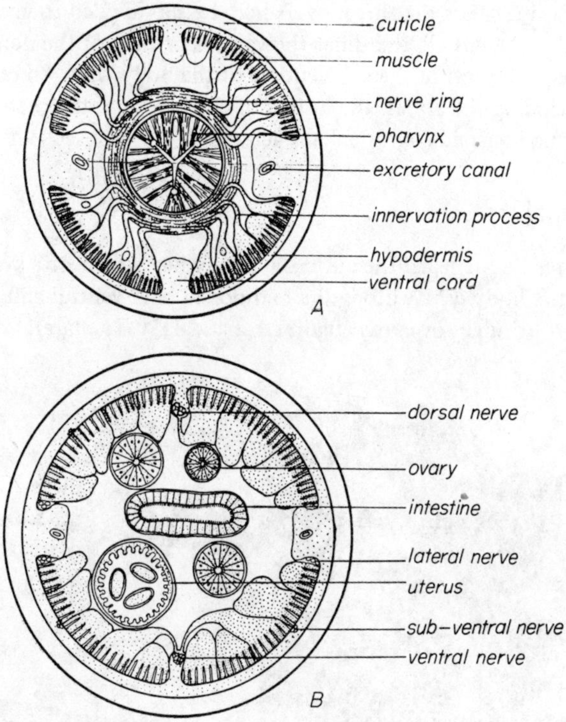

Figure 1.2 Transverse sections through the pharyngeal region (*A*) and the middle region (*B*) of a nematode.

immature adult and development of the gonads occurs. The adult nematode may continue to grow in size without further moults and in some examples (*Ascaris, Dracunculus*) they become very large.

1.2 The Organisation of the Nematode Body

1.2.1 Cuticle

Nematodes owe much of their success as a large and ubiquitous group to an organisation that includes a cuticle of varying complexity and considerable evolutionary plasticity. The cuticle forms the outer covering of the body but it also lines the pharynx, excretory pore, rectum, cloaca, spicule pouches, vulva, amphids and phasmids (sense organs). The stylet, teeth, hooks, spicules and certain additional copulatory structures such as the gubernaculum, are formed from toughened and hardened cuticle. The cuticle which covers the body surface is basically a three-layered structure and consists of cortical, median and basal layers (figure 2.1). This three-layered structure can be reduced to two in some species whereas in others the three layers may be sub-divided to give more layers (see figure 2.2). The cuticle that lines the various orifices of the nematode is usually simpler in structure than the cuticle of the body wall. All cuticular structures, including the stylet, the linings of the pharynx, the rectum, the amphids and the phasmids, are shed at each moult[4,5,6] (see section 2.2).

1.2.2 Hypodermis

The hypodermis lies beneath the cuticle and may be cellular or a syncytium. It projects into the body cavity along the mid-dorsal, mid-ventral and the lateral lines to form four ridges or cords (figures 1.2 and 1.3). The lateral hypodermal

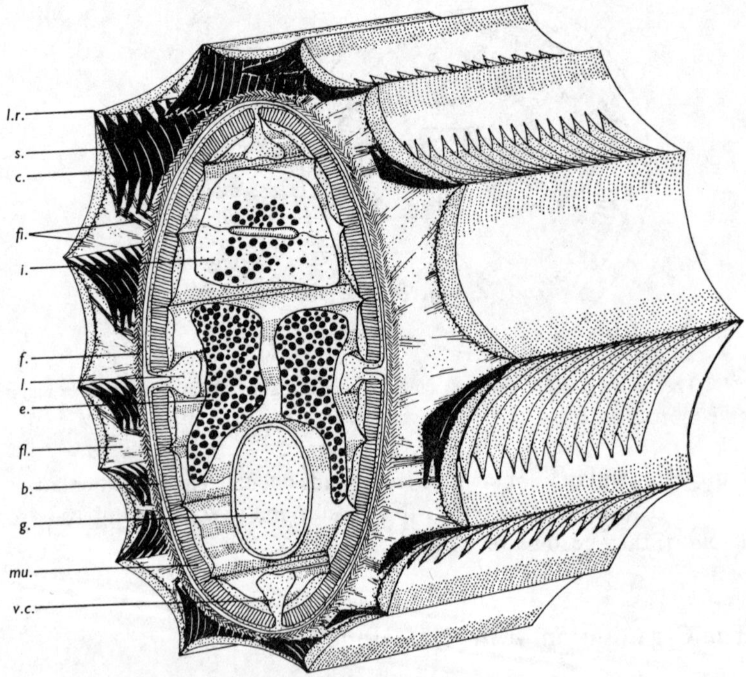

Figure 1.3 Diagram of a section taken from the middle region of *Nippostrongylus brasiliensis* (adult) to show the internal structures of the nematode and also the unusual cuticle [Lee, *Parasitology,* 55 (1965)]. *b* basal layer of cuticle; *c* cortical layer of cuticle; *e* subventral (excretory) gland; *f* fibres in basal layer of cuticle; *fi* fibrils; *fl* fluid-filled layer of cuticle; *g* gonad; *i* intestine; *l* lateral cord; *lr* longitudinal ridge of cuticle; *mu* muscle; *s* strut which supports the ridge of cuticle; *v.c* ventral cord.

cords are the largest and contain the excretory canals when these are present (figure 1.2). The nuclei of the hypodermis are found only in the cords. The hypodermis contains large amounts of reserve materials (lipid, glycogen) and haemoglobin is present in some animal-parasitic nematodes. Fibres traverse the hypodermis and appear to be concerned with the attachment of the muscles to the cuticle. Half-desmosomes form points of attachment between the hypodermis and the basal layer of the cuticle; fibres extend across the hypodermis from these half-desmosomes to desmosomes on the inner membrane of the hypodermis where they form points of attachment to the muscle cells (figure 1.4).

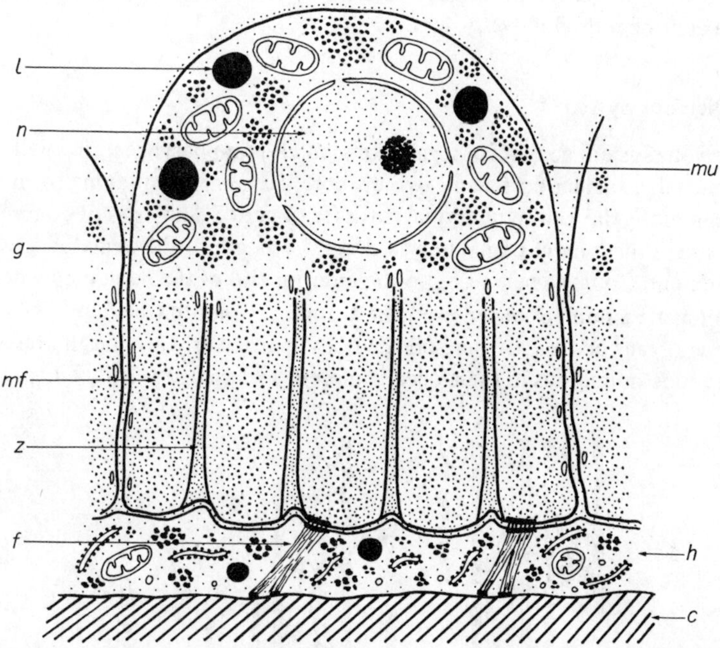

Figure 1.4 Transverse section through a muscle of the body wall to show the internal arrangement of the cell and its relationship with the hypodermis and cuticle. *c* cuticle; *f* fibres connecting muscle to cuticle; *g* glycogen; *h* hypodermis; *l* lipid; *mf* myofilaments; *mu* muscle; *n* nucleus; *z* z-band region of muscle.

1.2.3 Muscles

The muscles of the body wall of nematodes are unusual. They are all spindle-shaped, longitudinal muscles and the cells are divided into a contractile and a non-contractile portion (see figures 1.2 and 1.4). There are no circular muscles in the body wall. The muscles are obliquely striated (see figures 7.1 and 7.3) and the filaments of actin and myosin lie in the basal, contractile portion of the muscle cell (see section 7.1).[7] In some types of muscle the contractile elements also extend up the sides of the muscle cell (see figure 7.1). The remainder of the cell contains the nucleus, mitochondria, glycogen and fat. The muscles are not innervated by nerves extending to end plates on the surface of the muscle cell, as in most animals. Instead, innervation processes extend from the muscle to the neural tissue and make synaptic contact at the surface of the longitudinal nerves or the nerve ring (figures 1.2 and 7.2). In some nematodes the muscles in each section are interconnected with each other by cytoplasmic bridges and this may be important in muscle co-ordination.[4] The muscles of the body wall are attached to the hypodermis by desmosomes and exert their pull on the cuticle through fibres in the hypodermis (section 1.2.2). Not all muscle cells are in the body wall of the nematodes; there are specialised muscles associated with the alimentary system and the reproductive system.

1.2.4 Nervous System

The nervous system consists of a ganglionated circum-pharyngeal ring with a large ventral, a smaller dorsal and two or more lateral nerves running from this nerve ring along the four hypodermal cords (see figure 7.4). Series of commissures interconnect these longitudinal nerves along the length of the nematode. Several nerves extend from the nerve ring to the anterior end of the nematode where they innervate a number of sense organs around the mouth (see figure 7.4). A few sense organs occur elsewhere on the body, particularly around the tail where they are most numerous in the cloacal region of males (figure 1.5). The pharynx

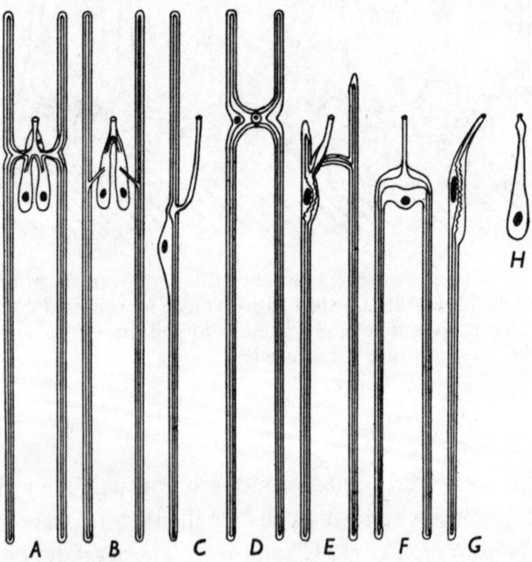

Figure 1.5 Representative types of excretory system found in nematodes. (*A*) Rhabditoid type, an 'H'-shaped system with two sub-ventral gland cells, the lateral canals are embedded in the lateral cords; found in *Rhabditis, Rhabdias* and the Strongylina in general. (*B*) Variant of (*A*), found in *Oesophagostomum.* (*C*) Tylenchoid type, an asymmetric system with the lateral canals and gland cell confined to one cord. (*D*) Oxyuroid type, an 'H'-shaped system without sub-ventral gland cells and with a shortened terminal duct. (*E*) Ascaroid type, a shortened 'H'-type which is almost an inverted 'U' in form. (*F*) Cephaloboid type. (*G*) Anisakid type, a reduced form related to *E* but with one lateral canal absent. (*H*) Single ventral cell type, present in Chromadorina, Monhysterina and Enoploidea. (After Chitwood and Chitwood.[10])

is also innervated throughout its length by a pharyngeal-sympathetic nervous system.

1.2.5 Pseudocoelom

The body cavity of nematodes is a pseudocoelom and is filled with fluid. The pseudocoelom is of fundamental importance as its fluid bathes the internal organs and also forms the hydrostatic skeleton of the nematode. Nutrients from the intestine, oxygen from the environment, and waste products have to cross the pseudocoelom to reach the various tissues and organs. It is not surprising, therefore, to find that the pseudocoelomic fluid is a very complex solution; that of *Ascaris*, for example, contains a variety of proteins, fats, carbohydrates, enzymes, nitrogenous compounds, inorganic ions and haemoglobin.[8]

The pseudocoelomic fluid and body contents are always under pressure, owing to the tonicity of the body-wall musculature, and this is of great significance in the locomotion, feeding, excretion and, indeed, in the structure of nematodes.[9,2]

1.2.6 Pseudocoelomocytes

The pseudocoelomocytes (also called coelomocytes, stellate cells, athrocytes, phagocytic or giant cells) are ovoid or many-branched cells lying in the pseudocoelom (figure 1.1) and are usually two, four or six in number. Various functions have been attributed to these structures. They may be phagocytic; in *Ascaris* the pseudocoelomocytes engulf bacteria injected into the body cavity, although ink particles are not phagocytosed. Another function ascribed to these structures is the storage of insoluble waste products, but there is little evidence in support of this. The uptake and concentration of dyes, such as methylene blue and neutral red, by these cells does not necessarily indicate an excretory function because many types of animal cell take up these stains. They may be absorptive in function and remove unwanted metabolites from the body fluid in some way, or they may manufacture essential components for the body fluid. It has also been suggested that they are an oxidative centre for the nematode, but this is unlikely. Early workers believed that the pseudocoelomocytes were in continuity with the excretory system but this has been shown to be incorrect. Their physiological role in nematodes remains uncertain.[10]

1.2.7 Alimentary System

There is a mouth, surrounded by lips, a buccal cavity of varying shape, a muscular and glandular pharynx, which usually has a triradiate lumen, an intestine, rectum and anus (figures 1.1, 1.2). The pharynx is a powerful pumping organ in most species and usually contains three glands that open into the lumen at varying positions along the length of the pharynx. The intestine is lined by microvilli and appears to be both secretory and absorptive in function. A valve or sphincter separates the intestine from the rectum or cloaca in most species. The alimentary system is described in greater detail in section 3.1.

1.2.8 Excretory System

The excretory system of nematodes was originally assigned this function on purely morphological grounds. It varies in structure (figure 1.6) and in some

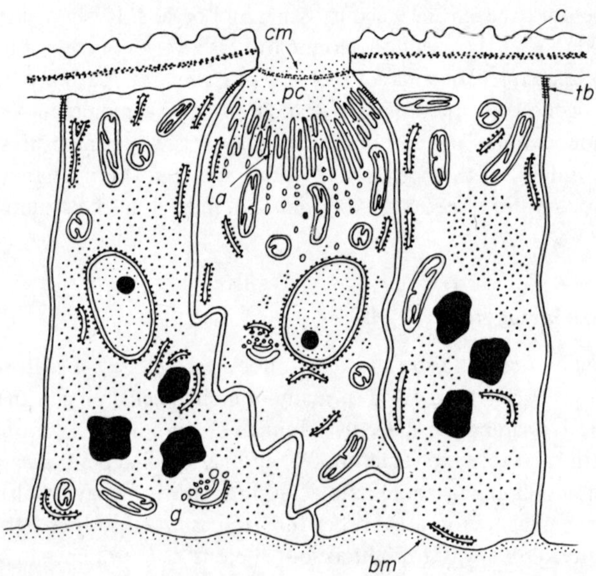

Figure 1.6 Diagram of a hypodermal gland cell and associated non-glandular cells of the lateral and ventral hypodermal cords of *Capillaria hepatica.* The hypodermal gland cell is situated below a pore in the cuticle and is surrounded by non-glandular cells. [After Wright, *J. Morph.*, 112 (1963).] *bm* basement membrane; *bl* boundary layer; *c* cuticle; *cm* cap material; *g* Golgi apparatus; *la* lamellar apparatus; *pc* pore chamber; *tb* terminal bar.

groups it appears to be completely absent. There are two basic types of so-called excretory system, glandular and tubular. The glandular system is found in many free-living nematodes. It consists of a ventral gland cell situated in the pseudocoelom near the base of the pharynx and usually has a terminal ampulla which opens to the exterior on the ventral surface by means of a pore (figures 1.1 and 1.6*H*). The tubular system varies in structure in different groups of nematodes. It is a simple H-system in the oxyurids with a canal in each lateral cord. These lateral canals unite by a short duct, which opens to the exterior through a common excretory duct and pore on the ventral surface near the base of the pharynx (figure 1.6*D*). Some nematodes (*Nippostrongylus*) also contain a pair of sub-ventral glands (figure 1.3) that open into the canal system just behind the excretory pore (figure 1.5*A* and *B*).[11] Other types of nematode possess a modified form of the canal system (figure 1.6*C, E, F* and *G*). In *Ascaris* and some other nematodes the canal system is in the form of an inverted 'U', the anterior limbs of the H-type system being absent (figure 1.6*E* and *F*).[10]

The lateral canals are intracellular and the whole canal system has only one nucleus. The duct of the canal lies within the elongated excretory cell and numerous vacuoles or canaliculi merge with it (figure.1.5). The glands associated with the canal system are often large (figure 1.3) and contain a number of enzymes in some animal-parasitic nematodes. They appear to have a secretory rather than an excretory function in these species (section 5.3).[11,12,13] There is a complete absence of flame cells in these systems but in certain nematodes pulsations of various parts of the system have been observed.

1.2.9 Reproductive System

The sexes are usually separate, the males frequently being smaller than the females. Males have one or two testes opening into a seminal vesicle and then into a vas deferens. This latter structure may be divided into glandular and ejaculatory parts. The vas deferens opens into a cloaca. Many male nematodes possess copulatory spicules (figures 1.1, 1.5, 6.1 and 6.2) that lie in cuticular-lined pouches connected to the cloaca. The spicules consist of hardened cuticle with a cytoplasmic core containing a nerve that runs to the tip of the spicule.[14] In some groups of nematode the area around the cloaca is expanded to form a copulatory bursa (see figure 6.2).

The spermatozoa of nematodes are amoeboid, but different types of spermatozoa characterise different groups of nematode. Some are round or ovoid whereas others have the superficial appearance of flagellated sperm but true flagella are not present (see figure 6.3 and section 6.1.1). Females have one or two ovaries, which open into an oviduct(s) and a uterus or uteri (see figure 6.4). The spermatozoa are usually stored in a seminal receptacle at the ovarian end of the uterus and are often closely associated with the wall of this region. Muscle fibres cover the uteri of some species and bring about movement of the contents. The uteri often end in an ovijector, which is very muscular and this, together with body movements and the high internal pressure of the body contents, serves to expel the eggs through the vagina. The eggs of nematodes vary in shape and size (see figure 6.5) but are essentially ovoid and have three main layers, the middle of which contains a chitin–protein complex. This is the only structure of nematodes that contains chitin. Some nematodes release sex attractants and in *Heterodera* they are known to be species-specific.[15,16,17]

The sex ratio is density-dependent in a few nematodes, and is possibly controlled by nutritional factors. Thus, in *Mermis* the eggs develop as females or males depending on the number that are eaten by the grasshopper, a preponderance of males occurring in heavily infected grasshoppers. Similarly, sex determination in *Meloidogyne incognita* apparently depends on the abundance of food, for the proportion of the sexes varies with the intensity of infestation of the host, with more males appearing in large populations. Female juveniles of *Meloidogyne* can undergo sex reversal; if the nutritive conditions are altered they develop into adult males with two testes (normal males have one testis).[15,18,19]

1.2.10 Hypodermal, Caudal and Rectal Glands

Hypodermal glands are present in the lateral cords of many marine nematodes. There are usually two sub-lateral rows of these unicellular glands and they are associated with pores on the surface of the nematode. Some species also possess a ventral row of these glands. Each pore on the surface of female *Chromadorina germanica* leads through a duct to one of these merocrine gland cells. A nerve cell is associated with each gland cell and a dendrite with a sensory cilium projects into the duct of the gland.[20] Gland-like cells occur in rows (called bacillary bands) along the length of the animal-parasitic Trichuroidea (*Trichuris, Capillaria*). These gland cells (figure 1.6) also open to the exterior through cuticular pores and some have sensory neurons associated with them.[4,21] The function of these hypodermal gland cells is not known.

Caudal gland cells (usually three) are present in many marine- and fresh-water nematodes. They produce a secretion that hardens in water and fixes the tail of the nematode to the sub-stratum. There is a high rate of production of this secretion as individuals can fix themselves to the sub-stratum many times in a short period of time. Ejaculation of the secretion is controlled by a spinneret apparatus. This consists of a retractor muscle attached to a small cone which blocks the aperture of the ducts at the tip of the tail. Retraction of the muscle withdraws the cone and ejects the secretion through the ducts and out of the aperture.[22] The head of the nematode can move in all directions while the tail remains fixed to the sub-stratum by means of this secretion and is of obvious value to a small organism that lives in rapidly moving currents of water.

Rectal glands lie in the pseudocoelom near the intestinal–rectal junction and open by ducts into the rectum or cloaca. There can be three to six of these glands depending on the species and sex of the nematode. The function of these glands is unknown in most nematodes. They discharge their secretion to the exterior through the rectum and anus and it could be that they are a source of sex attractant in some species. In adult female *Meloidogyne* they produce a gelatinous material, which surrounds and protects the extruded eggs at the posterior end of the swollen female.[4]

2 Cuticle, Moulting and Growth

2.1 The Cuticle

2.1.1 Structure

The cuticle of nematodes plays an important role in their physiology. It can be a simple or a complex structure, which varies from one genus to another and may differ in juveniles and adults of the same species. It is basically a three-layered structure and usually consists of cortical, median and basal layers (figure 2.1). This three-layered pattern occurs in the juvenile stages of most nematodes but in the adult cuticle modifications often occur (figures 1.3 and 2.2).[4,5,6,23]

In the typical three-layered cuticle (figure 2.1) the cortical layer is covered by a membrane-like structure (we shall call it the epicuticle) which varies in thickness from 25 to 40 μm; it appears to consist of an outer membrane, which is similar in appearance to a triple-layered unit membrane, a middle osmiophobic layer and an inner osmiophilic layer. There is controversy about the origin of this layer. The main cortical layer is single in most juveniles and in many free-living adults, but inner and outer cortical layers occur in many nematodes.

The median layer of the cuticle (also called the matrix or homogenous layer) is a relatively structureless layer, which is composed of numerous fine fibres in most nematodes. In the marine nematode *Acanthonchus duplicatus* the median

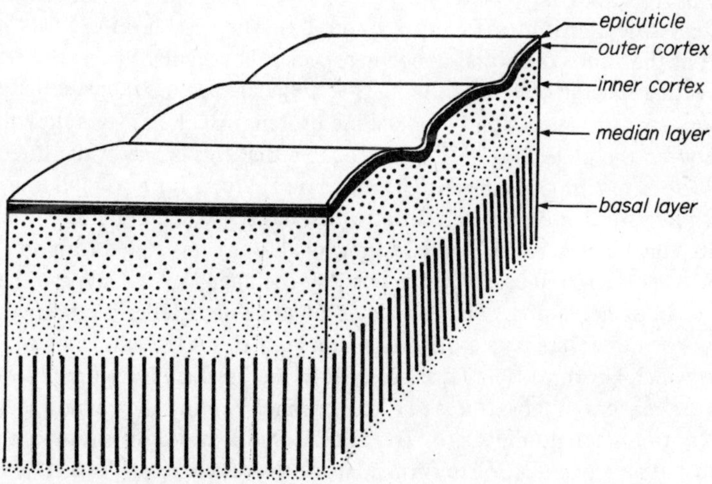

Figure 2.1 Diagram of a three-layered cuticle. This type of cuticle covers the body surface of many juvenile and adult nematodes. Modifications of the cuticle are thought to have arisen from this type.

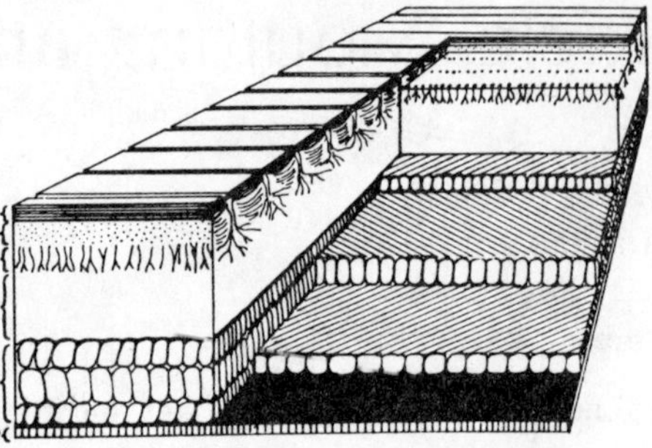

Figure 2.2 Diagram showing transverse, longitudinal and tangential sections of the cuticle of adult Ascaris. Note the sub-division of the basal layer into three fibre layers. [After Bird and Deutsch, *Parasitology*, 47 (1957).]

layer is traversed by dense material, which appears to be skeletal in function; pores extend from it into the hypodermis.[23] In adult *Nippostrongylus* this layer is filled with a red fluid containing haemoglobin. Skeletal struts or ribs lie in this fluid-filled layer and support the longitudinal ridges of the cuticle. Fibres, which exhibit the typical periodic cross-banding of collagen, traverse this fluid-filled middle layer and connect the basal layer to the cortical layer. They also suspend the skeletal struts in this layer (see figure 1.3).[5,24,25] The adult cuticle of several species of nematode can increase in thickness and in *Ascaris* most growth occurs in the median layer (see figure 2.9 and section 2.3).[4,5,6]

The basal layer varies in thickness from a very thin layer containing randomly orientated fibrils in some free-living nematodes (*Rhabditis pellio* and *Acanthonchus duplicatus*) to a thick layer containing larger fibres in a number of the larger, animal-parasitic nematodes (figures 2.1 and 2.2). The basal layer of many juveniles and of the adults of several free-living species, has regularly arranged striations with a spacing of about 20 nm. These regular spacings indicate that this layer consists of a tough, almost crystalline protein, which has very close linkages between the molecules.[4,6,26] It is probable that this is one of the three natural forms of fibrous collagen. The other two natural forms of collagen are an unstriated form and a striated form, which has a spacing of about 64 nm. The unstriated layers of nematode cuticle are probably the unstriated form of collagen.[27] This striated basal layer is possibly the main skeletal layer in the cuticle of these nematodes. The basal layer of some nematodes is modified to form two or three fibre layers. A good example of this is the cuticle of adult *Ascaris,* which has three fibre layers present in the basal layer (figure 2.2). Each fibre layer consists of large fibres that run in spirals around the nematode at about 75° to the longitudinal axis. The middle of the three layers crosses the other two at an angle of 135° to form a lattice that alters its angles as the nematode moves (see section 8.1).[2] Most of the larger nematodes that possess fibre layers have only two (*Oxyuris, Nippostrongylus*). These fibres are an important component of the cuticle in larger nematodes.[4,5,9]

The cuticles of the second stage juveniles and males of *Heterodera* species are similar to the basic plan described for juvenile nematodes (figure 2.1). The cuticle of gravid females differs from the basic pattern in ways that suit their swollen form, their sedentary habit and their eventual roles as protective cysts. The three layers of males and juveniles are supplemented by an extra fibrous layer in *H. schachtii* and *H. rostochiensis* and also by a fifth layer in *H. rostochiensis.* The fibres in this fifth layer are arranged helicoidally, as in the chitin of insect endocuticle, but they consist of collagen in the nematode. Chitin fibres in insect endocuticle lie in a matrix of cross-linked protein of low modulus and strength. This system has a large strength-to-weight ratio, good tensile, flexural and compressive strength and is much stronger than either of the two compounds alone. If fibres of material of high tensile strength and elasticity are orientated in a matrix of weak tensile strength and elasticity, the strength of the two-phase system is further increased. A helicoidal arrangement of the fibres gives an isotropic structure, but once it is formed it can no longer expand. The fibres in the fifth layer of *H. rostochiensis* fit this pattern and it has been suggested that such an isotropic layer would favour a spherical structure, in response to increased internal pressure as the female swells.[27] A somewhat similar structure is present in the innermost layer of the cuticle of female *Mermis nigrescens*[6] and may be formed in a similar manner to the endocuticle of insects and the innermost layer of female *Heterodera rostochiensis.* The cortical layer of adult female *Mermis nigrescens* cuticle is traversed by canals that appear to open by pores on to the surface of the nematode.[6]

Females of *Bradynema,* which live in the haemocoele of the mushroom fly, have no cuticle but the hypodermis has a microvillous surface and this forms the outer covering of the nematode.[28]

The cuticle may have characteristic longitudinal and transverse ridges and there may also be annulations, spines, punctations, or inflated areas. Most of these modifications involve the cortical layer only but, in some species, the median layer may also be involved. Lateral extensions of the cuticle, called lateral alae, are present in many nematodes. These may be stiffened, fin-like structures or blunt ridges of cuticle and may play an important role in locomotion. The fin-like alae are mainly found in animal-parasitic nematodes that are free-swimming in gut contents or other fluids and are topographically dorsal and ventral in position when the worm is swimming. They may help to stabilise the worm and may also allow greater thrust against the fluid environment during swimming, as occurs with the dorsal and ventral fins of eels. The blunt, lateral alae are found in most nematodes that live in soil or mud, including the free-living juveniles of animal-parasitic species. As most nematodes move on their sides, that is, with their lateral surfaces topographically dorsal and ventral, the blunt, lateral alae of a side will be against the sub-stratum. This will give a broad area of contact between the nematode and the sub-stratum, thus allowing greater purchase against the sub-stratum during movement.[24] Superficial annulations of the cuticle of nematodes (figure 2.1) allow easier bending of the nematode during locomotion. Adult *Nippostrongylus* possess longitudinal ridges in the cuticle (see figure 1.3) and they appear to assist coiling of this nematode. The ridges also abrade the intestinal mucosa of the host and may thus assist in the feeding process.[24] No doubt the various modifications of the cuticle of nematodes can all be related to a specific function or functions.

2.1.2 Composition

The basic component of the cuticle of nematodes is a form of collagen associated with hyaluronic acid, chondroitin sulphate, acid mucopolysaccharide, and small amounts of lipid. In vertebrate tropocollagens the basic, triple-helical pattern is developed through interaction of three separate polypeptide chains. The triple helix of collagen from *Ascaris* cuticle is different, as it is formed when sub-unit polypeptide chains, of molar mass 60 000, fold back upon themselves to form a collagen-type triple helix. These sub-units are held together by a disulphide bond.[29] Cuticular collagen of *Ascaris* contains amino acids that differ in their concentrations from those contained in vertebrate collagen, elastin, fibroin, resilin and keratin (table 2.1), but is digested by bacterial collagenase. Another structural protein called cuticlin is present in the cuticle of *Ascaris.* It contains large amounts of proline and alanine and relatively low amounts of glycine and basic amino acids; it does not give the characteristic X-ray diffraction pattern of collagen nor is it susceptible to collagenase. This cuticlin may be located in the cortical layers of the cuticle.[30] There is also some evidence that the outer cortical layer is a collagenous layer that has been modified by the presence of disulphide and sulphydryl groups to form a keratin.[4,5] Polyphenol quinone

Table 2.1 Comparison of cuticlin from the cuticle of *Ascaris* with other structural proteins (Fujimoto and Kanaya[30])

Amino acid	Cuticlin (*Ascaris*)	Collagen (*Ascaris* cuticle)	Elastin (bovine)	Fibroin (*Bombyx*)	Resilin (locust)	Keratin (wool)
	Residues/1000 total amino acids					
Glycine	150	274	330	445	376	85
Small[a]	317	346	441	739	487	146
Imino acids[b]	301	312	221	3	79	85
Basic[c]	36	82	88	9	45	104
Acidic[d]	146	136	117	23	152	173
Half-cystine	24	27	0	0	0	106
Hydroxyproline	0	16	98	0	0	0

[a] glycine + alanine
[b] proline + hydroxyproline
[c] lysine + hydroxylysine + histidine + arginine
[d] aspartic acid + glutamic acid

tanning occurs in the cuticle of swollen female *Heterodera rostochiensis.* The cuticle of the swollen female becomes tanned by the action of polyphenol oxidase to produce a tough, resistant cyst that encloses the eggs after the death of the female.[31] Esterase enzymes and haemoglobin have been detected in the cuticle of adult *Nippostrongylus*; esterase is present in the median layer of *Ascaris* cuticle; and polyphenol oxidase is present in the cuticle of *Heterodera.* ^{14}C-proline is incorporated into the cuticle of adult *Ascaris* and of juvenile *Nippostrongylus* and this, together with the prescence of enzymes demonstrates that the cuticle is metabolically active and not an inert covering.[4,5,32,33]

2.2 Moulting and Ecdysis

2.2.1 Introduction

Most nematodes moult four times during their development and certain stages in the process of moulting are apparently controlled by neuroendocrine secretions as in arthropods.

Moulting occurs in four main steps:

(1) The stimulus is received and brings about the discharge of neurosecretory material that initiates step 2.

(2) The old cuticle separates from the hypodermis.

(3) The new cuticle forms between the hypodermis and the old cuticle, with or without absorption of parts of the old cuticle.

(4) Rupture and ecdysis of the old cuticle, or its remains, occur allowing escape of the juvenile or young adult worm.

During moulting the cuticle that covers the body of the nematode, together with the cuticular linings of the amphids, phasmids, buccal cavity, pharynx, excretory pore, and the rectum are shed. During moulting in plant-parasitic tylenchids the basal part of the stylet dissolves and the head is disengaged from the anterior part of the old stylet, which is shed with the old cuticle.

2.2.2 The Stimulus

In most nematodes the identity of the stimulus that initiates moulting is not known. In free-living forms it is possible that growth to a certain size triggers moulting, or it could be controlled by a biological clock, such as the gradual loading, followed by sudden discharge, of neurosecretory cells. In some plant- and animal-parasitic nematodes it can be initiated, at least for one moult, when the juvenile moves from a free-living to a parasitic existence. For example, host root exudates stimulate the fourth-stage juvenile of *Paratylenchus nanus* to moult. Movement from one environment to another within the host may also initiate moulting (for example, migration of *Nippostrongylus* from the lungs to the intestine). The infective-stage juvenile of several animal-parasitic species and the third-stage juvenile (dauer juvenile) of some free-living nematodes (*Rhabditis dubia*, which lives in cow dung) retain the cuticle of the second-stage juvenile as a sheath; ingestion by a suitable host, in the case of the animal-parasitic species, or transfer by flies to freshly deposited dung, in the case of *R. dubia*, provides the stimulus for exsheathment and the continuation of development. This, however, is a stimulus that brings about ecdysis and does not initiate moulting (see section 2.2.5).[4,6,34,35]

2.2.3 Formation of the New Cuticle and Ecdysis

The changes that occur during moulting seem to vary from species to species and, as moulting has been studied in relatively few species, it is difficult to generalise. Some nematodes shed the old cuticle almost intact whereas others break down part of the old cuticle and shed only the outermost layer or layers.

During moulting the old cuticle separates from the outer hypodermal membrane and thus, indirectly, loosens its attachment to the muscles. This explains why some nematodes, such as *Aporcelaimellus*, become immobile prior to and during moulting.[36] Leucine aminopeptidase increases in concentration in the hypodermis of *Xiphinema index* between moults and reaches a peak just before moulting commences. The concentration of the enzyme falls once moulting begins and, simultaneously, the cuticle becomes loosened from the hypodermis.[37] This suggests that the enzyme is released into the area between the cuticle and the hypodermis and attacks the cuticle, thus bringing about its detachment from the hypodermis. This enzyme may also be present in the exsheathing fluid of *Haemonchus contortus* (see section 2.2.5).[35] In some nematodes (*Meloidogyne javanica*) the hypodermis becomes more granular and the amount of RNA increases at an early stage in the formation of the new cuticle; in others (the final moult of *Nippostrongylus*) this does not occur until later in the formation of the new cuticle. This is associated with the formation of material for the new cuticle.

Globular moulting bodies appear in the hypodermis of *Hemicycliophora arenaria* at the beginning of the moult and are thought to be associated with the formation of cuticular protein. The new cuticle is formed in a series of layers, which gradually increase in thickness and complexity. The edge of the new cuticle (the epicuticle) forms at the outer surface of the hypodermis as a thin, membrane-like layer. This epicuticle may be a condensation of new cuticular material or it may be the hypodermal membrane, which has become transformed and underlain by a new hypodermal membrane. The cuticle is laid down between this membrane-like epicuticle and the hypodermal membrane.[38]

In *Meloidogyne javanica* the newly formed outer cortex increases in thickness together with the new inner cortical layer and basal layer as shown in figure 2.3. The new cuticle is laid down in folds, allowing rapid elongation of the nematode after ecdysis. In this nematode the gap between the old cuticle and the new becomes filled with granular material and the inner layers of the old cuticle begin to break down (figure 2.3). The old cuticle is thought to be absorbed through regions in the newly formed cuticle where the internal cortex has not formed, until only the outer cortex is left and is eventually ecdysed.[4]

During the first moult of *Heterodera schachtii,* which occurs in the egg, strong contractions bring about detachment of the thin cuticle at both ends of the juvenile. This shed cuticle is then completely hydrolysed.[39] The first moult in *Ascaris* occurs in the egg and here also there is almost complete hydrolysis of the old cuticle.[4]

Cuticle formation of the body wall and of the stomodaeum are synchronised in *Aporcellaimellus,* the new stylet being formed at the same time as the new cuticle of the body wall. In this nematode, projections of the hypodermis

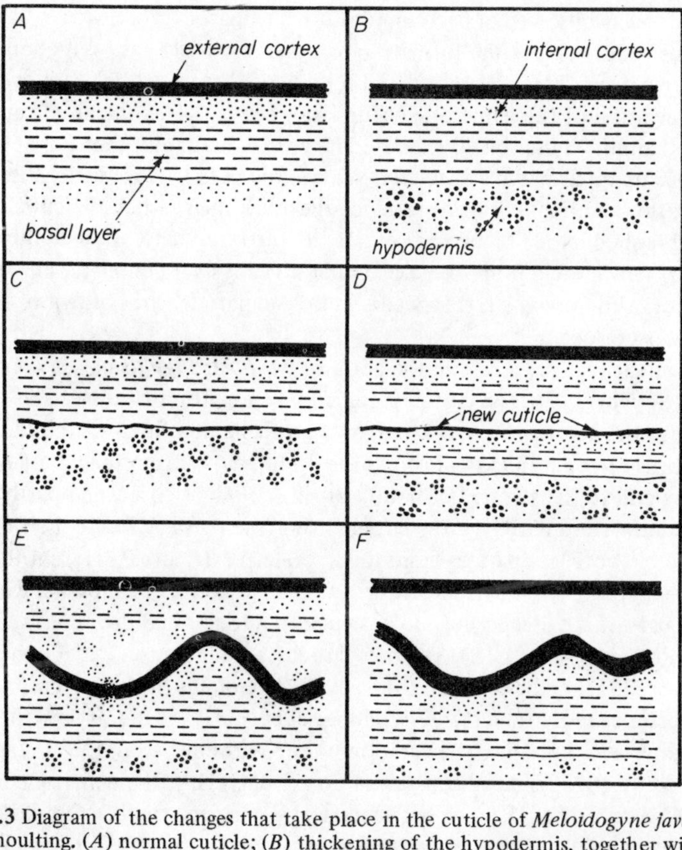

Figure 2.3 Diagram of the changes that take place in the cuticle of *Meloidogyne javanica* during moulting. (*A*) normal cuticle; (*B*) thickening of the hypodermis, together with an increase in the number of ribosomes, occurs; (*C*) the old cuticle has separated from the hypodermis and the new external cortical layer has begun to form on top of the hypodermis; (*D*) dissolution of the inner layers of the old cuticle has begun while the new cuticle begins to form the three basic layers and gradually increases in thickness; (*E*) inner layers of the old cuticle are increasingly dissolved (and resorbed?) while the new cuticle increases in thickness and becomes convoluted; (*F*) only the external cortical layer of the old cuticle remains and the new cuticle is almost fully formed. The old cuticle is then shed and the nematode extends itself, thus flattening the convolutions in the new cuticle. (After Bird.[4])

extend into the newly forming cuticle but retract from it later in the moult to leave a lacunar system in the basal fibre layers. Differentiation of the cuticle into distinct zones occurs once secretion of the new cuticle is completed. There is no dissolution of cuticle.[36]

The cuticle of *Hemicycliophora arenaria* is unusual as it consists of an outer cuticular sheath and a more normal cuticle. Both are formed when the nematode moults and should be considered as two parts of the same cuticle. Moulting commences with separation of the old cuticle from the hypodermis. The new sheath is then formed at the surface of the hypodermis. This is followed by formation of the new inner cuticle. The old sheath and inner cuticle then break down and may be absorbed by the nematode. Breakdown and reformation of the muscles was observed at each moult in this nematode.[38]

Ecdysis occurs before the formation of the new cuticle has been completed in *Turbatrix aceti.* Rapid secretion and folding of the new cuticle, followed by quick elongation of the nematode occurs after ecdysis. The old cuticle appears to be shed almost intact but the striated layer becomes amorphous in appearance.[6] During moulting in *Panagrellus silusiae* the cuticle is gradually discarded, except during the final moult to produce the adult male, when the cuticle splits and is shed in one or two pieces. At the final moult to produce the adult female the new cuticle folds extensively before ecdysis is complete, but this is the only occasion on which this occurs in this nematode. Breakdown of the old cuticle does not occur.[40]

The fourth-stage juvenile and the adult of *Nippostrongylus* have a complex cuticle (see figure 1.3). The cortex is folded into longitudinal ridges and struts extend into the apex of these ridges. These struts lie in a fluid-filled middle layer. Fibrils of collagen extend across the middle layer from the basal layer to the cortex and to the struts. Formation of this cuticle is controlled partly by the hypodermis and partly by the cuticle itself (see figure 2.5). At the final moult the old cuticle separates from the hypodermis (figure 2.4*A*), another membrane appears to form beneath the outer hypodermal membrane, which becomes part of the epicuticle, and cuticular material is laid down between these two membranes (figure 2.4*B*). The new cuticle is folded into regular annuli with two to three of these annuli for each single annulus of the old cuticle. The cuticle becomes more folded and hypodermal cytoplasm extends into these folds (figure 2.4*C* and *D*), but after a time they withdraw from the annulations leaving behind an 'M'-shaped ring of material, which girds the nematode in each annulus (figure 2.4*E*). The area between the 'M'-shaped structures and the hypodermis rapidly enlarges and develops into the two-fibre layers of the new cuticle (figure 2.4*F*). Once this has occurred there is rapid deposition of material above the 'M'-shaped structures and growth of this area of the cuticle (figure 2.4*G*). The 'M'-shaped structures, which seem to play a part in controlling the formation of the two parts of the cuticle, then gradually flatten out and disappear, while the outer part of the cuticle condenses into the cortex and the struts, leaving a fluid-filled cavity above the fibre layers (figure 2.4*H*). It is at this stage

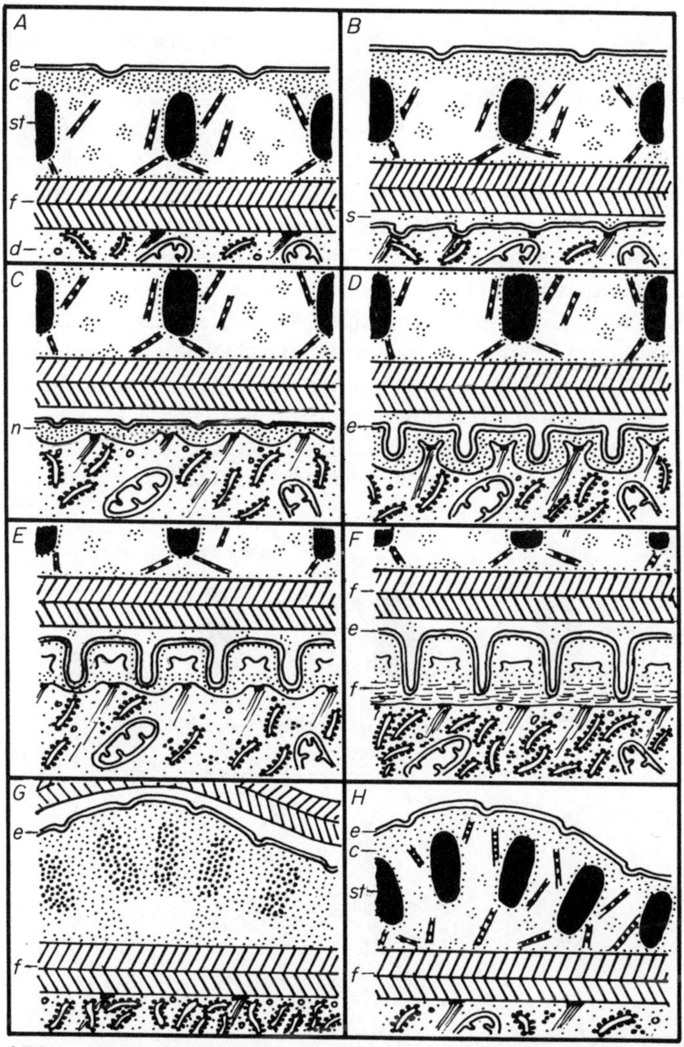

Figure 2.4 Diagram to show the stages in the formation of the adult cuticle during the fourth moult of *Nippostrongylus brasiliensis* (see figure 1.3 for a diagram of the cuticle). (*A*) and (*B*) separation of the juvenile cuticle from the hypodermis; (*B*) and (*C*) early stages in the formation of the adult cuticle, it is thought to be laid down between a new hypodermal membrane and the old hypodermal membrane which becomes part of the epicuticle; (*D*) the new cuticle becomes more extensively folded and the hypodermis extends into each fold; (*E*) the cuticle is much more extensively folded and formation of the basal (fibre) layer has begun, the hypodermis has begun to retract from the folds leaving 'M'-shaped structures in each fold, there is rapid increase in the amount of granular endoplasmic recticulum in the hypodermis at this stage; (*F*) later stage than (*E*) to show the fibre layers almost completely formed; (*G*) after the stage shown in (*F*) there is a rapid increase in thickness of the outer part of the cuticle, the 'M'-shaped structures stay close to the fibre layers and seem to play an important part in separating the fibre layers from the rest of the cuticle, the struts are shown at an early stage in their formation and the fluid-filled layer has begun to appear; (*H*) the adult cuticle shortly after ecdysis, the new cuticle rapidly elongates causing the folds of the superficial annulations to flatten, the struts are in the final stages of formation, collagen fibres extend from the outer to the inner layers of the cuticle and to the struts, there is reduction in the amount of granular endoplasmic reticulum in the hypodermis. There is no dissolution of the old cuticle which is shed almost intact. (After Lee.[25])

c cortex; *e* epicuticle; *f* fibre layers; *h* hypodermis; *n* new cuticle; *s* space between old and new cuticles; *st* strut.

that the old cuticle is shed more or less intact. The ridges are formed when the hypodermis pushes into the space between the hypodermis and the old cuticle, deposits new cuticle and then retreats back to the normal position. The muscles of the body wall are attached to the inner hypodermal membrane at desmosome-type junctions and fibres run from these to anchor at half-desmosome-type thickenings at the hypodermis–cuticle junction. These half-desmosomes separate from the old cuticle when moulting commences but very quickly anchor to the new cuticle–hypodermal membrane (figure 2.4*A–D*). Thus the muscles can exert leverage on the new cuticle very soon after the hold on the old cuticle has been lost.[25] Control of the formation of the various layers and of the struts in the cuticle of *Nippostrongylus* poses an intriguing biophysical problem which remains to be solved but formation of the collagen may be similar to that in fibroblasts (figure 2.5). Moulting in *Nematospiroides dubius* appears to be very similar to that in *Nippostrongylus.*[41]

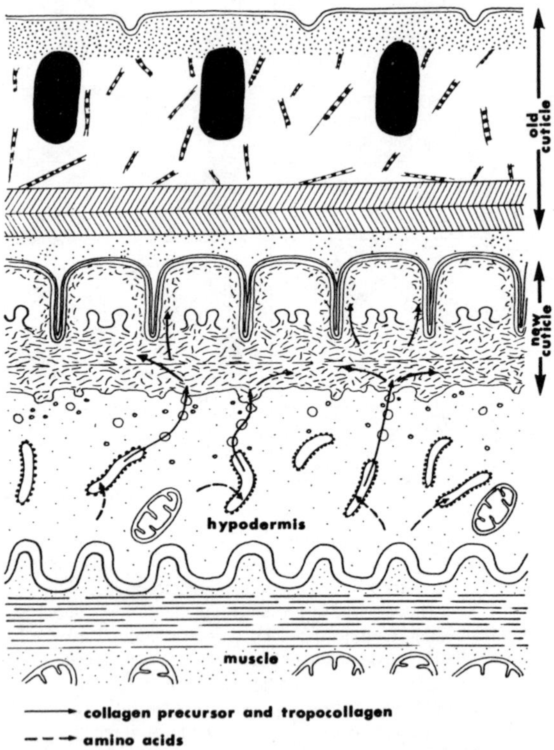

Figure 2.5 Diagram to explain how the new cuticle of *Nippostrongylus brasiliensis* (see figure 2.4) may be formed from amino acids (based on fibroblast production of collagen). Amino acids enter the granular endoplasmic reticulum in the hypodermis and are made into collagen precursors on polyribosomes. The collagen precursor is then carried to the outer hypodermal membrane in small vesicles and released into the new cuticle. In the cuticle the collagen precursor or tropocollagen becomes transported to, and orientated in, the various layers of the cuticle. (After Lee.[25])

Several other nematodes do not dissolve the old cuticle during moulting; these include *Turbatrix aceti, Panagrellus silusiae, Aphelenchus avenae, Hirschmaniella gracilis, Aporcellaimellus* spp.[38,40] On the other hand, several nematodes do dissolve the cuticle during moulting. The inner layers of the moulted cuticle of *Trichinella spiralis* undergo a gradual breakdown during moulting.[42] The basal lamella degenerates first, followed by the striated layer, but the cortical layers are not affected. It is not known if this degeneration is brought about by moulting fluid secreted by the nematode or by digestive enzymes of this host. Nor is it known if the dissolved, cuticular material is absorbed by the nematode. The reasons for these different methods of moulting and ecdysis probably lie in the different environments of the different species. For example, the first-stage juvenile of *Ascaris* moults in the egg, where space and nutrients are at a premium. It is to the advantage of the juvenile to reduce the thickness of the cast cuticle as this will allow it more space to develop within the egg. The egg is a closed system and the developing juvenile relies entirely upon its food reserves until it reaches another host; it is, therefore, to its advantage to recycle the proteins obtained from absorption of the old cuticle. Similarly, *Meloidogyne javanica* moults after it has attached itself to the root of the host plant. Reduction in thickness of the old cuticle will make ecdysis and growth easier for this semi-sessile nematode. On the other hand, the fourth-stage juvenile and adult of *Nippostrongylus* live between the villi of the host's intestine and are liable to be carried out of the host in a stream of gut contents if they are unable to maintain their position in the mucosa. It is, therefore, to the advantage of the fourth-stage juvenile to moult as quickly as possible and dissolution of the old cuticle could be a hindrance in this respect. The nematode also has a plentiful supply of food and so can afford to discard the proteins of the old cuticle. The later juvenile stages of *Trichinella* also live against the mucosa of the intestine, but breakdown of the cast cuticle does occur in this species.[42] Nematodes that live a relatively free existence, either in the soil or in the alimentary tract of other animals, can wriggle free from their old cuticle much more easily than those nematodes that live in confined environments. It is, therefore, not surprising to find that some of these 'free' nematodes cast most of the old cuticle intact, whereas nematodes in confined environments tend to breakdown and may absorb most of the old cuticle. Obviously more examples of moulting and ecdysis in nematodes from different environments are needed before any generalisations can be accepted.

2.2.4 Neurosecretory Control of Moulting and Ecdysis

There is a correlation between neurosecretory activity and the process of moulting in some nematodes, but most of the neurosecretory activity that has been studied is associated with ecdysis. Ecdysis of the fourth-stage cuticle in *Phocanema decipens* is accompanied by synthesis of leucine aminopeptidase in the excretory gland and release of the enzyme through the excretory pore into the gap between the old and the new cuticle. The enzyme then attacks the old cuticle, which eventually ruptures to free the newly moulted nematode. Synthesis and release of this leucine aminopeptidase is under the control of neurosecretory cells in the ventral ganglion. When the fourth-stage juvenile is grown in a nutrient medium the neurosecretory cells produce and release this enzyme and ecdysis ensues. If the juvenile is kept in a non-nutrient medium a new cuticle forms but the neurosecretory cells remain inactive, enzyme is not produced and ecdysis does not occur.[43,44]

Insect juvenile hormones, or their analogues, affect development of some nematodes. Farnesol and farnesyl methyl ether (FME) can inhibit moulting, prevent the development of male copulatory appendages, induce multiple moults, or prevent ecdysis, depending upon the concentrations used and the species or stage of nematode tested. Pronasterone and ecdysone have no effect upon a variety of nematodes, including *Caenorhabditis briggsae, Aphelenchus avenae* and *Strongyloides fülleborni*, but ecdysone and a substance that has been isolated from freeze-dried *Haemonchus contortus,* and which has similar activity to ecdysone when injected into insects, inhibited the growth of free-living stages of *H. contortus.* A compound with juvenile hormone activity (when tested in insects) has been isolated from *Haemonchus contortus* infective juveniles. It is similar to FME when injected into insects and when run in thin-layer chromatographs[45]. Farnesol, FME and this substance from *H. contortus* inhibit growth of *Panagrellus redivivus* and *H. contortus*, and delay maturation of *C. briggsae.* FME and a synthetic preparation of insect juvenile hormone, when applied to the fourth-stage juveniles of *Phocanema in vitro*, stimulate the neuroendocrine system of the nematode. This subsequently brings about production and release of the moulting fluid required for ecdysis.[46]

2.2.5 Ecdysis of Ensheathed Juveniles

Moulting of the infective stages of several animal-parasitic nematodes (for example, *Trichostrongylus, Haemonchus, Ancylostoma*) and of free-living nematodes that produce a resistant or dormant 'dauer' juvenile, is incomplete. The first- and second-stage juveniles of these species feed on bacteria in the faeces or in the soil. The third-stage juvenile does not feed and is entirely enclosed by the cuticles of the second-stage juvenile in which even the old anal, oral and excretory pore orifices are closed. These juveniles thus possess the new cuticle of the third-stage juvenile and the moulted, but not ecdysed, cuticle of the second-stage juvenile. Ensheathed juveniles are similar to infective eggs of *Ascaris* in that they have a lowered rate of metabolism. The term exsheathment is used to describe ecdysis of these ensheathed juveniles. Exsheathment of several species seems to require a specific external stimulus. This comes from fresh dung in the case of dauer juveniles (they are carried to it on the legs of insects) or from the alimentary tract of a specific host in the case of infective juveniles of trichostrongyles such as *Haemonchus* and *Trichostrongylus.*[34,35]

High concentrations of dissolved gaseous carbon dioxide and/or undissociated carbonic acid at 37 °C (conditions that are normally found in the alimentary tract of the host) are the main stimuli that initiate exsheathment (ecdysis) of *Haemonchus* and *Trichostrongylus* but other factors such as the pH and the prescence of reducing agents are also important (figure 2.6).

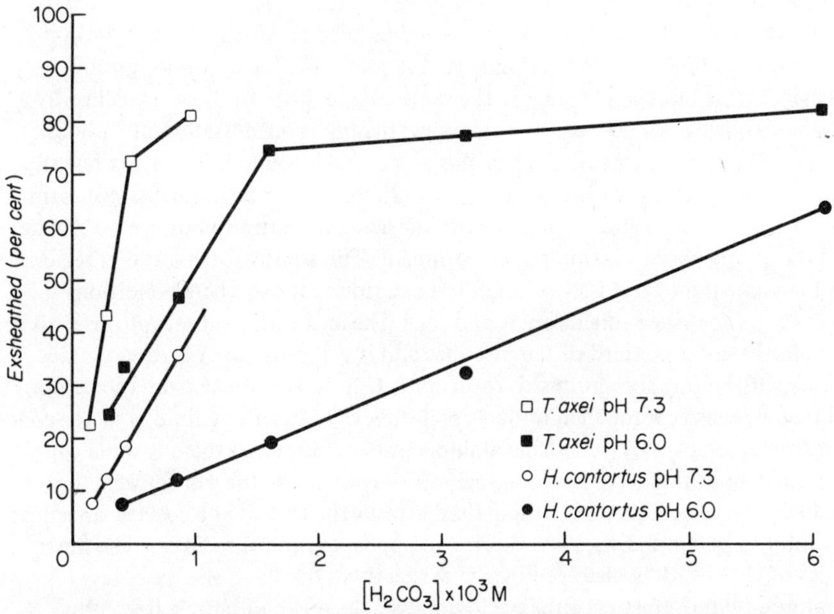

Figure 2.6 The effect of different concentrations of undissociated carbonic acid [H_2CO_3] plus dissolved gaseous carbon dioxide and a reducing agent on the exsheathment of juveniles of *Trichostrongylus axei* and of *Haemonchus contortus* at different pHs. (After Rogers.[48])

Host enzymes are not responsible for exsheathment of *Haemonchus contortus* and *Trichostrongylus axei* juveniles. This was shown by enclosing infective juveniles in a cellophane sac and placing them in the rumen of sheep. Although digestive enzymes of the host were unable to pass through the cellophane, digestion of the sheath and release of the cuticular cap occurred as in normal exsheathment. Fluid taken from exsheathing juveniles also brings about exsheathment of unstimulated juveniles. Infective juveniles of *Dictyocaulus viviparus* and *Trichostrongylus colubriformis* appear to require pepsin from the host, as well as carbon dioxide, to bring about exsheathment and infective-stage juveniles of *Nematodirus battus* require exposure to high pH. Bile and bile salts potentiate exsheathment of some species.[35,47]

In *Haemonchus* and *T. axei* the whole process of exsheathment takes about three hours but the action of the stimulus is completed after fifteen minutes. Juveniles continue to exsheath in water at 38 °C after they are removed from a fifteen-to thirty-minute exposure to the stimulus.[35]

Exsheathment of juveniles of *T. axei* and of *Haemonchus contortus* increases as the concentration of undissociated carbonic acid increases (figure 2.6). These juveniles require much higher concentrations than juveniles of other nematodes examined and are the only species in which the process is not inhibited by a high concentration of carbon dioxide. *T. axei* normally exsheaths in the rumen and maximum exsheathment takes place between pH 7 and 8 when the concentration of dissolved, gaseous carbon dioxide and undissociated carbonic acid is above 0.5×10^{-3} M. Juveniles of *Haemonchus* need about three times the concentration of undissociated carbonic acid required by *T. axei* to bring about exsheathment but the pH range is the same (figure 2.6). This may explain why *Haemonchus* has such a narrow host range; the high concentration of carbonic acid necessary for exsheathment of this nematode is found in relatively few situations other than the rumen of ruminants. *T. colubriformis* normally exsheaths in the acid stomach and requires a carbonic acid concentration of 5×10^{-3} M at pH 4 to bring about maximum exsheathment. The adult worms settle in regions of the alimentary tract that are slightly posterior to those where conditions necessary for exsheathment are found.[35,48] The exsheathment receptor is situated in the anterior third of the juveniles and it is thought that carbon dioxide reacts with sulphydryl groups to form an -S-C-S- link in the activated receptor.[35] This activated receptor then initiates secretion of exsheathing fluid, a major constituent of which may be leucine aminopeptidase (although there is some controversy about this),[35,49] from the excretory system into the gap between the old and new cuticles. This exsheathing fluid attacks the sheath at a specific area near its anterior end and causes a weak ring to appear around the sheath. The inner layer of the sheath is dissolved by the exsheathing fluid and the outer layer is ruptured by movements of the enclosed juvenile. A cap of cuticle becomes detached and the juvenile wriggles free (figure 2.7).

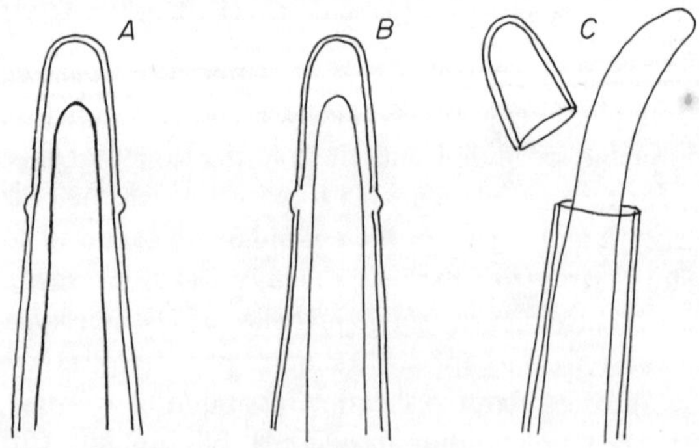

Figure 2.7 Stages in the exsheathment of infective juveniles of trichostrongyle nematodes (*Trichostrongylus, Haemonchus*). (*A*) the sheath, enclosing the third-stage juvenile, swells locally near the anterior end to form a weak ring in the cuticle; (*B*) the inner layers are digested by an exsheathing fluid which is thought to issue from the excretory pore; (*C*) the cuticle eventually ruptures around this weak ring and the anterior end of the sheath comes off as a cap to release the juvenile. [After Lapage, *Parasitology*, 27 (1935).]

The processes involved in exsheathment are probably not the same as those that initiate moulting. Moulting has already taken place in ensheathed juveniles and exsheathment, or ecdysis, only brings about the release of the juvenile from its sheath. These juveniles have apparently lost the ability to initiate ecdysis at the second moult and the stimulus must be supplied by a suitable host or, in the case of dauer juveniles, by a suitable environment. This appears to be a good adaptation to ensure survival of the species.

2.3 Growth

The growth of nematodes is a more or less discontinuous process as they undergo four moults during development. Growth may be continuous throughout moults, or a moult may be associated with a period of lethargus during which the nematode stops growing before, during, and just after, moulting (figure 2.8). The period between moults, and even after the final moult in some species such as *Ascaris*, is a period when extensive and rapid growth may occur. The duration of the lethargus differs in different species. For example, *Ascaridia galli* continues to

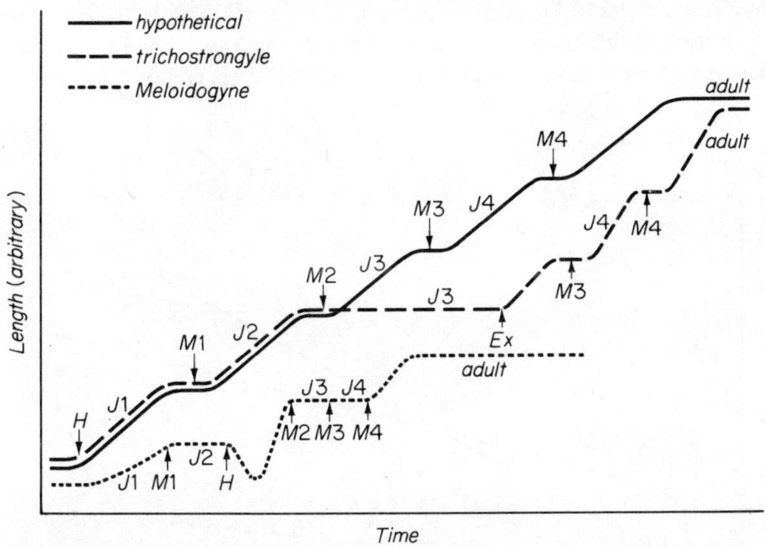

Figure 2.8 Growth curves during the life cycle of three types of nematode. The hypothetical curve shows well-marked periods of growth and lethargus and is probably typical of many free-living nematodes. The trichostrongyle curve shows a period of arrested development after the second moult and development proceeds after exsheathment. The growth curve of *Meloidogyne* shows an initial decrease in size of the second-stage juvenile after hatching, followed by a period of rapid growth, growth then ceases during the next three moults and continues after the last moult. *Ex* exsheathment occurs; *H* hatching occurs; *J* 1–4 juvenile stages; *M* 1–4 moult occurs.

grow without any sign of a lethargus while some other animal-parasitic nematodes, such as *Angiostrongylus cantonensis,* have well-marked periods of lethargus associated with certain moults. The second-stage juvenile of the plant-parasitic nematode *Meloidogyne javanica* grows quite quickly on the roots of the host until the onset of the second moult. It then undergoes a prolonged lethargus during which three successive moults occur; growth is resumed after the fourth moult (figure 2.8).[4,35]

At certain stages of their life cycle some nematodes grow more in width than in length (female *Tetrameres americana* and female *Meloidogyne javanica*), whereas some may stop increasing in length but their internal organs, notably the reproductive system, continues to grow. Some juvenile nematodes may be shorter after moulting than before. This occurs for an initial period of three days in the second-stage juvenile of *Meloidogyne* and is followed by a period of rapid growth (figure 2.8).[4]

The cuticle of nematodes acts as an exoskeleton and, if the nematode is to increase in size, the cuticle must either stretch, grow, or be replaced at a moult. Stretching, and thus thinning, of the cuticle is extremely unlikely to occur in most nematodes as this would weaken it. The cuticle of adult *Ascaris* increases in thickness as it grows in length and diameter, this change in thickness being directly proportional to the change in length of the nematode. All layers of the cuticle increase in thickness; the median layer increases more rapidly than the fibre layers and these more rapidly than the cortical layers (figure 2.9). This

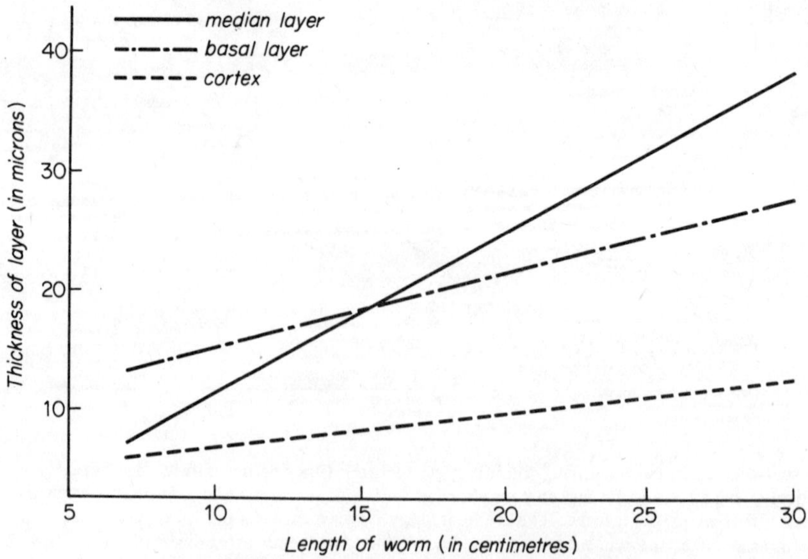

Figure 2.9 Increases in thickness of the cortical, median and basal (fibre) layers of the cuticle with increases in length of *Ascaris.* [After Watson, *Quart. J. Microsc. Sci.,* 106 (1965).]

means that more cuticle is being laid down as the nematode grows, presumably by the insertion of more protocollagen or tropocollagen into it from the hypodermis (see section 2.2.3). Some enzymes have been detected in the cuticle but it is not known whether these enzymes help to produce new cuticular matter, reorientate existing material, or serve some other purpose.[6]

Growth of the internal structure of the nematode is apparently less complicated than in many other animals. The cell number of most tissues and organs remains constant and growth occurs by an increase in size or change of shape of these existing cells. It is possible that some cells are destroyed and replaced to form new structures, for example, change in shape of the pharynx or buccal cavity from one stage to the next. The muscles of *Hemicycliophora arenaria* are known to break down and be reformed at each moult.[38] Cells of the intestine of some nematodes increase in number in the adult stage. There is also evidence that some of these cells are shed and replaced. There is an increase in the number of muscle cells in the body wall during development in some species; however this may not be a true increase but may be a sub-division of the existing cells as they grow larger. It is known that in several species of nematode the muscles of each of the four sectors are interconnected by cytoplasmic bridges. This may result from the splitting of muscle cells into smaller units, which remain in cytoplasmic continuity by means of these bridges. Most growth of the internal organs of immature adult nematodes occurs in the gonads.[4]

Nematodes can increase in size between moults and after the final moult. It has also been shown that the cuticle of at least some nematodes can increase in thickness after the final moult. Therefore the explanation that nematodes moult because they have an inert exoskeleton, which must be shed to allow growth to occur, would appear to be incorrect. It has been suggested that moulting is not necessary to permit rapid changes in size of the nematode but may be necessary to permit rapid changes in thickness of the cuticle.[9] This may apply to juveniles but adult *Ascaris* increase the thickness of their cuticle quite markedly. Moulting allows different stages of the nematode to modify existing structures, such as the buccal cavity and the pharynx,[50] but these are probably not the sole reasons for moulting. It must be stated that we are still unable to put forward a satisfactory theory as to why nematodes moult.

3 Feeding and Digestive Physiology

Nematodes feed upon a wide variety of foods (there are saprophagous, phytophagous, predatory, and parasitic species) but most individual species are restricted to one type of nutrition at any particular stage of development. The structure and physiology of the digestive system varies from species to species and also from juvenile stage to adult in some species. These differences can usually be related to different methods of feeding and to different types of food.

3.1 The Alimentary Canal

Most nematodes take their food into an alimentary tract, which is fundamentally the same in all groups (see figure 1.1). It is a straight tube from terminal mouth to sub-terminal anus and is divided into three parts:

(1) The *stomodaeum*, which includes the lips, mouth or stoma, buccal cavity and pharynx (oesophagus).

(2) The *intestine*, which can be divided into anterior (ventricular), middle, and pre-rectal regions.

(3) The *proctodaeum*, which consists of the rectum and anus in juvenile stages and females, or the cloaca with its associated structures in males. The stomodaeum and the proctodaeum are lined with cuticle that is shed at each moult.

The high turgor pressure in the fluid-filled pseudocoelom of most nematodes (about 70 mm Hg in *Ascaris*) has an important bearing on the structural organisation of the alimentary system. The intestine is normally collapsed, except when full of food, and the pharynx is a powerful sucking and pumping organ, which forces food into the intestine against this high turgor pressure.[2,9]

3.1.1 Stomodaeum

The greatest structural diversity in the alimentary system occurs in the stomodaeum and this is usually associated with the varied feeding habits of nematodes. The basic plan of the head is a mouth or stoma, which is usually circular and surrounded by six lips bearing sense organs (see figure 9.1). Modifications of this plan do occur; the lips may fuse to form three, as in *Ascaris* or *Heterakis* (figure 3.1*B*) or to form a ring around the mouth in the form of a sucker, as in *Syngamus* (figure 3.2*G*) or *Bunostomum* (figure 3.1*A*). The lips may form long processes that possess many spines (*Acrobeles*), or they may be small, tooth-like projections.

The buccal cavity lies between the mouth and the pharynx and varies greatly

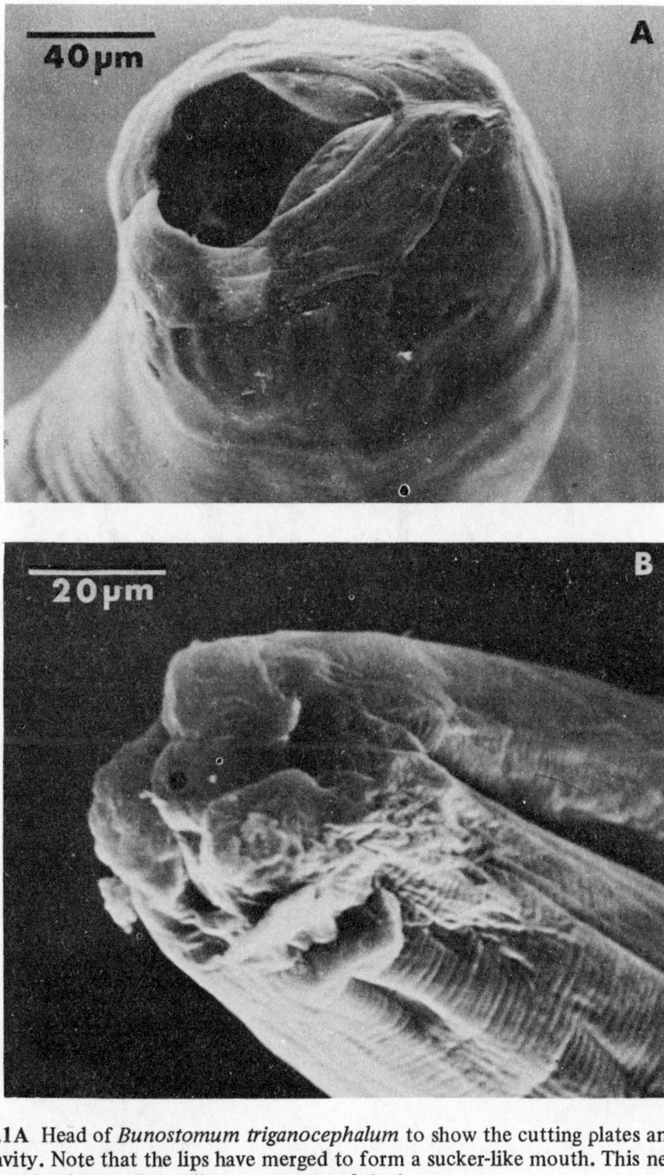

Figure 3.1A Head of *Bunostomum triganocephalum* to show the cutting plates and the large buccal cavity. Note that the lips have merged to form a sucker-like mouth. This nematode feeds upon the tissues of the alimentary tract of the host.

Figure 3.1B Head of *Heterakis gallinarum* to show the small mouth and the three lips. This nematode feeds upon the contents of the alimentary tract of the host.

in shape (figure 3.2). The simplest buccal cavity appears to be a cylindrical, narrow tube lined with cuticle and is usually associated with bacterial or fluid feeding, as in *Rhabditis* (figure 3.2*A*), adult *Ascaris*, and the juvenile stages of many animal-parasitic nematodes. The buccal cavity has been modified in several

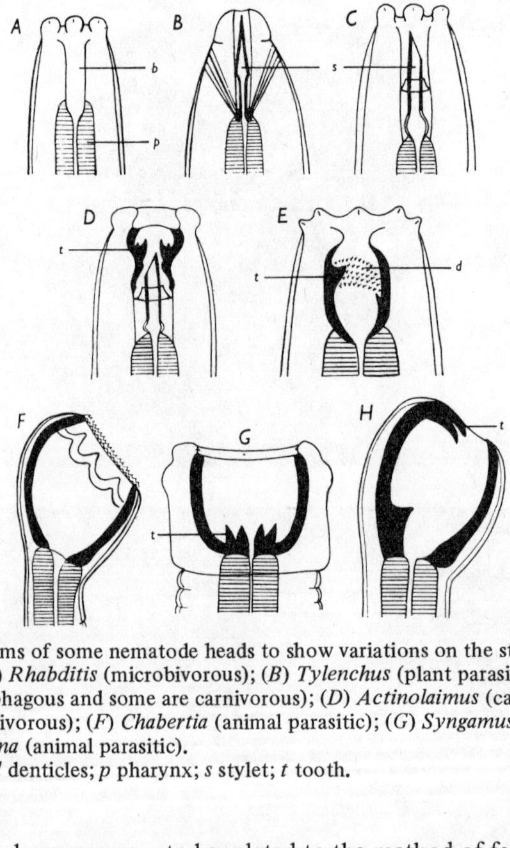

Figure 3.2 Diagrams of some nematode heads to show variations on the structure of the buccal cavity. (*A*) *Rhabditis* (microbivorous); (*B*) *Tylenchus* (plant parasitic); (*C*) *Dorylaimus* (some are phytophagous and some are carnivorous); (*D*) *Actinolaimus* (carnivorous); (*E*) *Mononchus* (carnivorous); (*F*) *Chabertia* (animal parasitic); (*G*) *Syngamus* (animal parasitic); *Ancylostoma* (animal parasitic).
b buccal cavity; *d* denticles; *p* pharynx; *s* stylet; *t* tooth.

ways and these changes appear to be related to the method of feeding of the species concerned. In many predatory nematodes (*Mononchus*) and in some animal-parasitic species (*Ancylostoma, Bunostomum, Syngamus*) the buccal cavity has become enlarged and cup-shaped. The cuticular lining of these nematodes is thickened, hardened, and shaped into small denticles, larger teeth, or cutting plates (figure 3.2*E, G* and *H* and figure 3.1*A*). A hollow stylet is present in the buccal cavity of most plant-parasitic nematodes (figure 3.2*B* and *C* and figure 1.5) and of some predatory species (*Actinolaimus, Dorylaimus*) (figure 3.2*C* and *D*). These stylets can be driven into prey or plant cells (figure 1.5) and the fluid contents withdrawn through the hollow centre of the stylet. Some stylets are solid, cuticular rods with a pointed end that can be driven into prey or a host cell to produce a hole from which fluid can be sucked through the mouth.

The stylet is derived from the cuticular wall of the buccal cavity (stomato stylet) in the Tylenchida but in some Dorylaimida the stylet develops in the pharynx (odonto stylet) and migrates into position in the buccal cavity after each moult. The stylet is moved in and out by means of specialised muscles during feeding. The lumen of the stylet of tylenchids is less than 0.5 μm in diameter and can prevent the entry of large particles. Secretions from the pharyngeal glands of some species pass through the stylet into the tissues of the prey, or into the cell of the plant, and contain enzymes that partially digest the cell contents before they are withdrawn through the stylet by the pumping action of the pharynx. In some predatory species the secretions may contain a paralysant. Secretions from some plant-parasitic species induce physiological changes in the plant and result in the formation of giant cells (section 3.2.2).[51] Some animal-parasitic nematodes (*Trichuris, Capillaria*) have a cuticular, spear-like structure (usually called the stylet although it is not the same as the stylet found in plant-parasitic nematodes) in their buccal cavity. In *Capillaria hepatica* the stylet is not associated with any muscles and cannot be protruded from the mouth but protrusion of the stylet does occur in *Trichuris* and it may be used during feeding.[52]

Muscles extend from the body wall to insert on the cuticular lining of the buccal cavity in many nematodes and when they contract they bring about expansion of the lumen, thus sucking food through the mouth into the buccal cavity prior to entry into the lumen of the pharynx. These muscles bring about biting movements in those species that possess teeth or cutting plates in the buccal cavity.

The buccal cavity opens into the pharynx (usually referred to in the literature as the oesophagus); this is a glandular pumping organ, which usually has a tri-radiate lumen. In some genera the median and posterior parts of the pharynx are swollen to form muscular bulbs, in others only a posterior bulb occurs, and some genera have none (figure 3.3). The cuticle that lines the lumen of the pharynx is not continuous with that which lines the buccal cavity. The cuticular lining of the posterior pharyngeal bulb may be modified to form three corrugated plates or flaps, which may play a role in moving food from the pharynx into the intestine and may also act as a grinding mechanism (figures 3.3 and 3.5). The three radii of the lumen dilate at their tips to form circular channels in many species (figure 3.3). The structure of the cuticular lining of these channels suggests that it gives rigidity to the channels. In some species the cuticle of the radii adjacent to these channels forms projections that interdigitate when the lumen is closed (figure 3.4) and thus prevent food particules entering the channels from the main lumen of the pharynx, but allow fluid to enter. This results in a bolus of food particles passing backwards along the central lumen of the pharynx while the fluid probably passes forwards through the semi-circular canals and is discharged from the mouth.[53]

The pharynx of most nematodes contains muscle cells, gland cells, supporting cells and nerve cells (figure 3.4) but it may be partially syncytial in some species

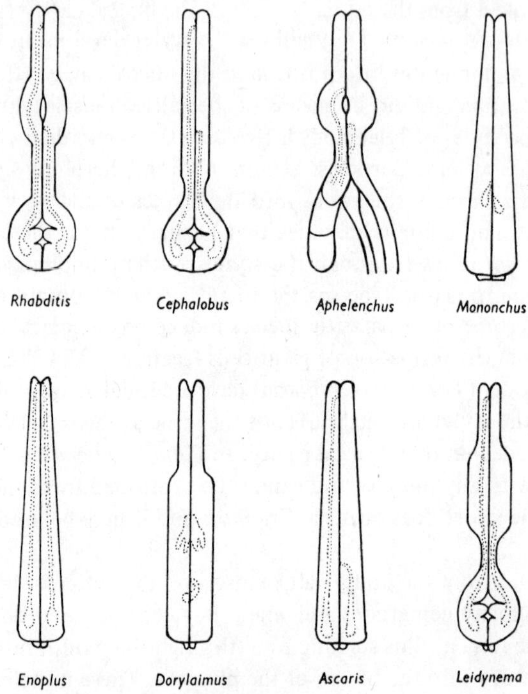

Figure 3.3 Diagrams of some nematode pharynges. The positions of the pharyngeal glands and their ducts are shown by broken lines. (Partly after Jones, *Plant Nematology* (H.M.S.O., 1959), and partly after Chitwood and Chitwood.[10])

(*Ascaris*). The muscles of the pharynx are radially arranged in groups along the length of the pharynx and upon contraction they pull on the cuticular lining, thus dilating the lumen (figure 3.4).[54] These muscles appear to be cross-striated and the total muscle length corresponds with one sarcomere in *Enoplus* and *Caenorhabditis*.[55,56] There is also a network of muscle fibres on the outside of the pharynx in a few species (*Trichinella, Capillaria*).[57,58]

Marginal cells are present at the apex of each arm of the triradiate lumen and contain fibres that appear to have a skeletal or supporting function as they extend from the cuticular lining of the lumen at the apex of each arm to the outer edge of the cell (figure 3.4).

Three nerves run along the length of each of the three sectors of the pharynx. In some nematodes they appear to be deeply embedded in the pharyngeal tissues, but they are still connected to the pseudocoelom by a narrow channel comparable to a mesaxon. A nerve ring is present in the posterior part of the pharynx. This pharyngeal nervous system is connected to the main nervous sytsem of the body at the anterior end of the pharynx (see section 7.2.1).[59]

There are usually three uninucleate glands in the pharynx, one in the dorsal sector and one in each of the two sub-ventral sectors. The glands may be multinucleate (Enoplida) and more than three glands may be present (*Longidorus* has five). In most species the dorsal gland opens near the anterior end of the pharynx and the two sub-ventral glands open into the lumen in the posterior part, and all three glands open into the buccal cavity of some species (in *Pontonema vulgaris* the glands open through three teeth at the base of the buccal cavity)[60] or into the middle or posterior pharyngeal lumen in others (figure 3.3). The duct of the dorsal gland may dilate into an ampulla at its distal end and serves as a reservoir for the secretion produced by the gland. The ampulla in adult *Nippostrongylus* opens through a duct into the mouth and allows the secretion to be passed directly onto the tissues of the host.[24] Secretory granules move along the glands and pass into the lumen of the duct when the glands are active. The ducts of the glands are lined with thin cuticle, except at the proximal end where they divide into many smaller channels, which are lined by the membrane of the gland cell.

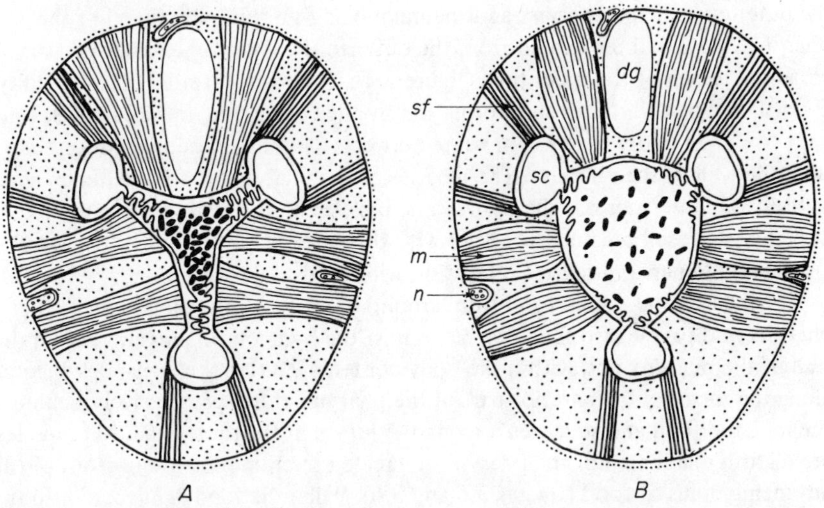

Figure 3.4 Transverse sections through the pharynx of *Heterakis gallinarum* with the lumen closed (*A*) and open (*B*). Note how the food particles (bacteria) remain in the central part of the lumen and do not enter the semi-circular canals. Fluid is squeezed into the canals when the lumen closes. *dg* dorsal gland; *m* muscle; *n* pharyngeal nerve; *sc* semi-circular canal; *sf* supporting fibres.

It is here that the granules pass from the cytoplasm of the gland into the lumen of the duct. The gland cells may vary in structure and composition from species to species and from juvenile stage to adult.[4] There may also be differences between the dorsal and the sub-ventral glands, for example, the dorsal and one of the sub-ventral glands of *Pontonema vulgaris* are acidophilic while the other sub-ventral gland is basophilic. One type of gland cell may produce an enzyme precursor and the other an activator, the two secretions yielding an active enzyme when they are mixed in the gut of the nematode.[60] These differences in different species, or throughout the life cycle of a single species, can be correlated with varying functions. One type of saliva may be produced during hatching, whereas others may be associated with penetration of a host, with extra-corporeal digestion or with differences in the type of food utilised.

In *Trichinella, Trichuris* and *Capillaria* there are no pharyngeal glands as described for other nematodes, but a series of gland cells (stichocytes) lie in the pseudocoelom and open into the lumen of the pharynx by means of ducts that pass through the wall of the pharynx. Although the exact function of these stichocytes is not known they are thought to produce secretions which may assist the passage of food along the pharynx or may contain digestive enzymes.[57,58]

3.1.2 The Functioning of the Pharynx as a Pump

The pharynx can be regarded as two tubes of low rigidity, the outer tube being the outer edge of the pharynx and the inner being the cuticular lining of the pharynx. Muscle fibres extend from the outer to the inner tubes and open the triradiate lumen of the inner tube (figure 3.4). The outer layer has a low rigidity and will not be able to resist deformation by the contracting muscles unless it is kept stretched by hydrostatic pressure between the outer and inner tubes, that is, in the wall of the pharynx. The muscles of the pharynx dilate the lumen and, at the same time, increase the hydrostatic pressure in the wall of the pharynx. It is this increased pressure, together with the unstable position of the walls of the dilated lumen and muscle relaxation which closes the lumen.

The muscles must suck food from around the mouth into the lumen of the pharynx and pass it into the intestine against the hydrostatic pressure within the wall of the pharynx and within the body contents. During the transport of food along the pharynx the anterior part of the pharynx closes progressively behind the ingested food and squeezes it along the lumen at the same time that muscles are dilating the lumen of the pharynx to create a suction pressure in front of the advancing bolus of food (figures 3.5 and 3.6). When the food is injected into the intestine the lumen of the pharynx is in open connection with the lumen of the intestine where the pressure will be equal to or slightly higher than the pressure in the body cavity of the nematode. The pharynx has to overcome this pressure when passing food into the intestine. If the pressure within the wall of the pharynx is less than or equal to the pressure in the lumen of the intestine then food will not pass from the posterior lumen of the pharynx into the intestine. Apparently the hydrostatic pressure created in the wall of the pharynx when the muscles of the pharynx contract is greater than that in the intestine. If successive waves of contraction carry more than one bolus of food backwards then the whole process will be under greater pressure than when a single wave of contraction is present and more efficient injection of food into the intestine will occur.[61]

Many plant-parasitic and mycophagous nematodes have short pumping regions in their pharynx (*Ditylenchus, Aphelenchoides*). In *Ditylenchus* the mouth stylet opens into the lumen of the anterior part (procorpus) of the pharynx, which is circular in section, lined with thick cuticle, is not occluded and is about 0.08 μm in radius. In the pumping region of the pharynx (the metacorpus) the lumen is triradiate when closed and circular when dilated by the pharyngeal muscles. The radius of the isthmus lumen is about 0.23 μm when dilated. Food is drawn into the pharynx by dilation of the metacorpus lumen and the lumen just behind the pump is closed. When the lumen of the pump closes the lumen of the isthmus is forced open and food passes quickly along it. The lumen of the third part of the metacorpus functions as an outlet valve for the pump. The pump can operate with an outlet valve and no inlet valve because the pump lies

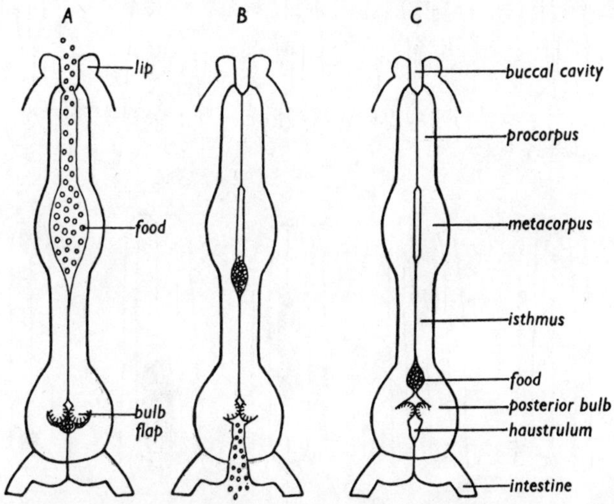

Figure 3.5 The structure and function of a rhabditid-type pharynx during feeding. Food particles, small enough to pass through the buccal cavity, are drawn into the lumen of the metacorpus by sudden dilation of the pro- and metacorpus (*A*). Closure of the lumen of the pharynx in these regions expels excess water (*B*–see also figure 3.3) and the mass of food particles is passed backwards along the isthmus by a wave of contraction of the radial muscles of the isthmus (*B* and *C*). Food is drawn between the bulb flaps of the posterior bulb by dilation of the haustrulum, which inverts the bulb flaps, and is passed to the intestine by closure of the pharyngeal–intestinal 'valve' (*B* and *C*). The bulb flaps contribute to the closure of the lumen in the bulb and may also crush food particles when they invert. [Diagrams from a description given by Doncaster, *Nematologica*, 8 (1962).]

between the procorpus, which has a lumen diameter of 0.88 μm, and the isthmus plus postcorpus, which has a lumen diameter of 0.23 μm when dilated. The resistance of the anterior pharynx to fluid flow will be sixty-eight times greater than that of the posterior pharynx. Thus when the dilated lumen of the metacorpus closes, sixty-eight times as much fluid will flow backwards into the posterior pharynx than forwards into the procorpus. Both inlet and outlet valves are associated with the metacorpus pump of *Hemicyliophora arenaria* and in *Xiphinema* a sheath of muscle, which surrounds the pharyngeal bulb, may provide an additional means of closing the long pharynx of this nematode.[62]

The pharynx of some nematodes has a posterior bulb and this seems to assist food into the intestine by pushing it out of the pharyngeal lumen. The three flaps of the bulb roll backwards against each other and appear to carry food into the intestine (figure 3.5).

The functioning of the pharynges of a *Rhabditis*-type nematode, of *Panagrellus*, of *Aplectana* and of *Ascaris* are shown in figures 3.5 and 3.6.[54,62]

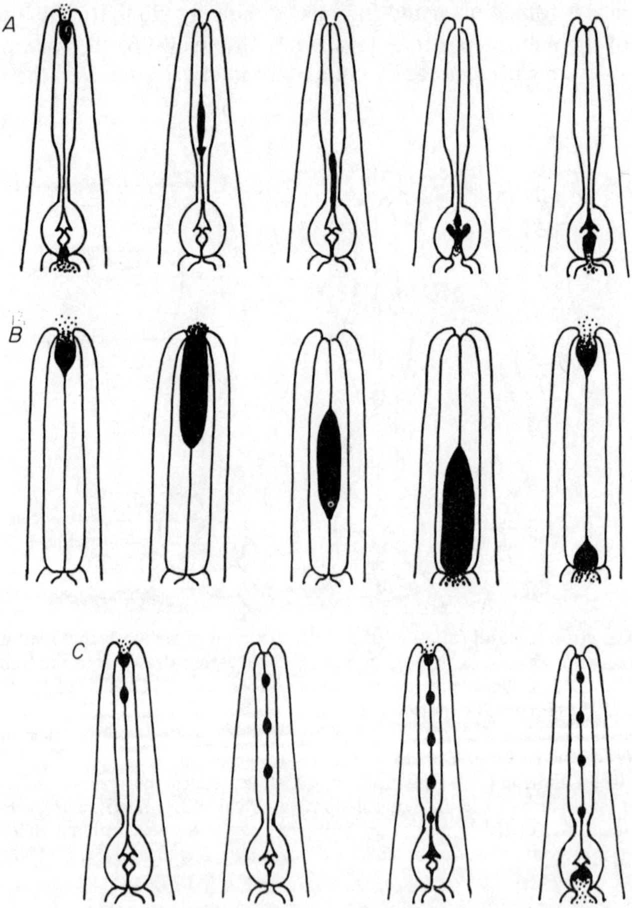

Figure 3.6 Variations in the methods of pharyngeal pumping. (*A*) *Panagrellus*; (*B*) *Ascaris*; (*C*) *Aplectana*. (After Mapes[54] and Crofton.[189])

3.1.3 Intestine

The intestine can be divided into three regions in some nematodes: the anterior (ventricular) region, the mid-region, and the posterior (pre-rectal) region. These regions are not usually distinctly separated but sometimes they differ from each other in the shape of the lumen, the height and contents of the cells and in their function. In most nematodes the intestine consists of a single layer of cells that enclose the lumen (figure 3.7) but in a few nematodes (*Ancylostoma*) the whole or part of the intestinal wall forms a syncytium. In some genera the intestine has no lumen and acts as a food storage organ (female *Meloidogyne, Mermis*). The

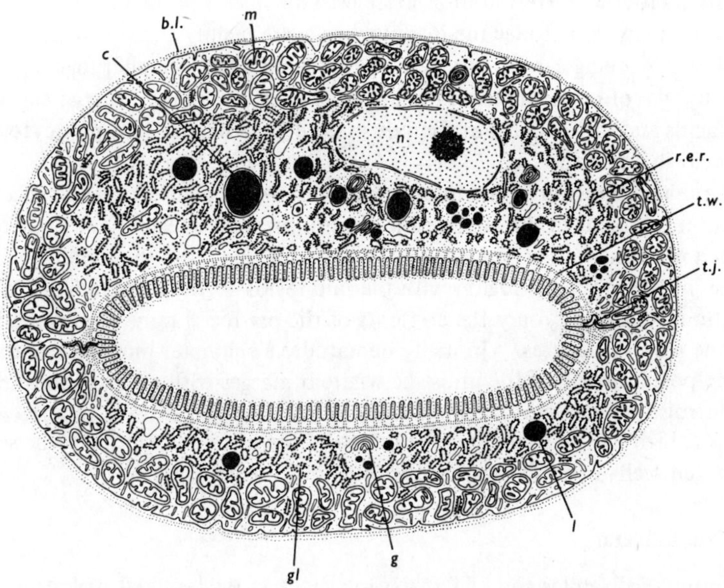

Figure 3.7 Transverse section through the anterior intestine of *Nippostrongylus brasiliensis* to show the cell contents. Note that the lumen is enclosed by two cells in any one section in this nematode (after Lee, 1969, *Parasitology*, 59). *b.l.* basal lamella; *c* crystalline inclusion; *g* Golgi apparatus; *gl* glycogen; *l* lipid; *m* mitochondrion; *n* nucleus; *r.e.r.* rough endoplasmic reticulum; *t.w.* terminal web; *t.j.* tight junction.

intestine is enclosed by a basal lamella, which varies in thickness with the species. Some nematodes (*Aspiculuris, Heterakis*) have a network of modified, somatic muscle fibres over the outer surface of the intestine for the whole of its length, or over the pre-rectal region, but this is not common.[10] The basal membrane of the intestinal cells may be smooth or extensively folded and it is probably very important in the transport of materials from the intestine to the pseudocoelomic fluid and to the other tissues of the body. Mitochondria are usually scattered within the cell but in some species (*Nippostrongylus*) they are concentrated in the basal region of the cell and form a broad band of mitochondria at the periphery of the intestine (figure 3.7). It is possible that this concentration of mitochondria is related to active transport of substances across the basal cell membrane. The cells contain the usual organelles associated with intestinal cells of some nematodes and are of unknown function. The cell membrane that lines the lumen of the intestine is extended into microvilli, which vary in appearance from species to species. The glycocalyx, which coats these microvilli appears to be important in the digestive process as it provides a large surface area upon which enzymes become absorbed. The main function of the microvilli is probably to increase the absorptive surface of the cell. The cells of the intestine can be both secretory and absorptive in function, for example, the intestinal cells of *Pontonema vulgaris* pass through an absorptive and then a secretory cycle.[60] Similarly merocrine secretion of esterase enzyme occurs in the intestine of *Ascaris*. In many nematodes the intestinal cells are mainly absorptive in function, most digestion being carried out by enzymes secreted by the pharyngeal glands.[4]

The uptake of nutrients from the intestine is mainly in the form of sugars, amino acids and fatty acids but there is some slight evidence that pinocytosis may occur in a few species.

The pre-rectal region of the intestine appears to be involved in the process of defecation as well as in the absorption of food. In some species (*Aphelenchoides*) modified intestinal cells form valve-like structures that close the lumen at the junction of the pre-rectal region with the mid-region of the intestine, during defecation. As a result, only the contents of the pre-rectal lumen are voided when the worm defecates.[63] In many nematodes a sphincter muscle seals the extreme posterior end of the intestine where it merges with the rectum or cloaca and controls the release of the intestinal contents at defecation.[10] In *Aphelenchoides*, which has no sphincter, the lumen of the intestine appears to be sealed by transient cell–cell adhesions at the intestinal–rectal junction.[64]

3.1.4 Proctodaeum

The rectum of juveniles and of females is a short, cuticular-lined structure. In males the intestine and the vas deferens open into a cloaca which is lined with cuticle; the cloaca contains the copulatory spicules and other accessory copulatory structures, when present. The rectum and the cloaca open to the exterior through a slit-like anus which has no anal sphincter and which appears to be kept closed, but not sealed, by the pressure of the body fluid. Muscles extend from the wall of the anus to the body wall and dilate the anus at defecation. Some nematodes (*Mermis*) have no anus and the intestine ends blindly.

Three to six rectal glands open into the rectum of most nematodes. The usual function of these glands is unknown but in female *Meloidogyne* they produce a gelatinous matrix which is extruded through the anus as a result of rhythmic contractions of the anal muscles. This gelatinous matrix surrounds the eggs that have been produced by the female. It is probable that in *Meloidogyne* these glands have become specialised for this purpose (see section 1.2.10).[4]

3.2 Feeding

3.2.1 Microbivorous and Saprophagous

Bacterial- and fluid-feeding nematodes usually have a simple alimentary system, with little modification of the mouth parts (*Rhabditis*) (figure 3.2*A*). A few have small teeth at the base of the buccal cavity (*Pontonema*) and in others the oral region forms elaborate processes (prolabae) which possess many bristles (*Acrobeles*). Many nematodes (for example *Rhabditis*) inhabit the soil where they feed on bacteria or decaying matter, which they draw into the alimentary tract by means of the sucking action of the pharynx (figures 3.5 and 3.6). *Turbatrix aceti* feeds in a similar manner upon bacteria and yeasts in malt vinegar. *Acrobeles* and related genera also feed upon bacteria, but the sieve-like bristles of the prolabae restrict the size of the particles that the nematode ingests. *Pontonema vulgaris*, which feeds unselectively on the mud in its environment, digests bacteria and some organic materials in the alimentary tract, but small inorganic particles and indigestible organic material are passed out in the faeces. During feeding the pharyngeal-intestinal valve closes, the radial muscles of the pharynx contract, the dilated lumen fills with food and the ingested food mixes with secretions from the three pharyngeal glands, which open into the buccal cavity. The anterior end of the pharynx then closes, the pharyngeal–intestinal valve opens, the muscles of the pharynx relax as the lumen closes and the food is forced into the lumen of the intestine. This process is repeated about once a second and the whole intestine can fill with food in five minutes[60]. Feeding in a *Rhabditis*-type nematode, in *Panagrellus* and in *Aplectana* is shown in figures 3.5 and 3.6.

Some nematodes have hardened and ridged cuticular plates in the posterior pharyngeal bulb and these may crush larger particles as they are pushed backwards into the intestine by the rolling action of the plates (figure 3.5). The nature of the food and the feeding mechanisms of the free-living juvenile stages of many animal-parasitic nematodes are essentially the same as in free-living nematodes such as *Rhabditis*; for example, the first- and second-stage juveniles of *Ancylostoma* and of *Nippostrongylus* feed upon bacteria in the faeces of the host. Third-stage juveniles of these species do not feed but exist on their food reserves until they infect another host, or die when their reserves are exhausted.[65]

3.2.2 Phytophagous and Mycophagous

Phytophagous nematodes feed upon diatoms, algae and cells of higher plants and mycophagous species feed upon fungal hyphae. They have a stylet with which they pierce the plant cell and which may also be used to withdraw the contents of the cell (figures 3.2*B* and *C* and 3.8).

Many of the stylet-bearing nematodes attack living cells of higher plants, either as ectoparasites, which feed externally upon the cells of the plant (*Longidorus, Xiphinema*) (figure 1.5), or as endoparasites, which migrate into and feed upon internal cells and tissues of the plant (*Heterodera, Ditylenchus, Radopholus*)[50] (see figure 3.8).

Behaviour leading to feeding or penetration falls into three sequences in most tylenchid species.[66] The nematode moves randomly in the soil, with periods of activity interspersed with quiescent periods, until a host is located. When it detects an attractive chemical stimulus from a plant it moves up the concentra-

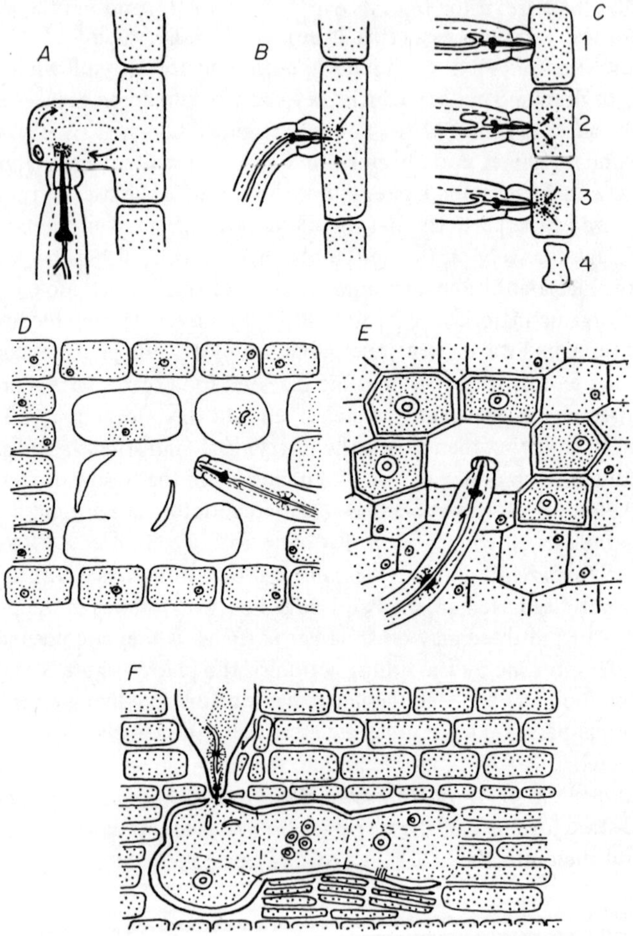

Figure 3.8 The methods of feeding of several types of phytophagous and mycophagous nematodes and their effects on plant cells. (*A*) Cyclosis continues during prolonged feeding. No visible destructive effect on the cell. (*B*) Immediate removal of cell contents through the stylet. (*C*) Delayed removal of cell contents. (1) Contact made with cell; (2) penetration of the cell wall occurs, cyclosis stops, saliva is injected into the cell and the cytoplasm becomes granular (0–40 min); (3) ingestion of cell contents occurs (40–90 min); (4) the stylet is withdrawn and the cell collapses. (*D*) The nematode secretes enzymes into the tissues, cells near the nematode enlarge and separate, cytoplasm withdraws from the cell walls and the walls collapse, cavities form in the tissues as the cells lyse. (*E*) Nurse cells of *Tylenchulus.* Nuclei and nucleoli of cells around the head of the nematode enlarge, the cytoplasm becomes dense, the cell walls thicken and the cells become nurse cells upon when the nematode feeds. (*F*) Syncytium of *Heterodera* (*Meloidogyne* form rather similar syncytia). Nuclei of the affected cells enlarge, cell walls dissolve and the cytoplasm of several cells merges into one unit; the cytoplasm becomes dense and granular and the remaining cell walls thicken. These enlarged cells appear to be a form of transfer cell. The nematode feeds on these giant cells. (After Dropkin.[51])

tion gradient towards the source (see section 9.3.1). This migration is accompanied by searching movements of the head, exploratory probing with the stylet, and the accumulation of secretory material in the ducts of the pharyngeal glands. This is called widespread exploration. When the nematode reaches the surface of the fungal hypha or the plant, it glides over the surface making intermittent probes with the stylet and occasionally stopping to explore the cell surface more intensively. This is called local exploration and probably represents a search for a suitable spot to attack. The sensory lips (see section 9.1.2) are pressed against the surface of the plant cell and rubbed against it, with brief pauses when probing with the stylet occurs, during exploration of the cell surface. The nematodes seem to prefer rigid to flexible cell surfaces as most probing occurs here. The mouth and stylet are placed at right angles to the cell surface when a suitable site has been found while the rest of the body usually remains immobile and is often braced against surrounding soil particles or projections on the plant or hyphal surface. This allows maximum thrust to occur. The stylet is then repeatedly everted against the cell wall until the cell is punctured; this is carried out more frequently, more rhythmically, and more powerfully than exploratory probings. The nematode leaves and continues widespread exploration if the cell contents are not suitable but ingestion or penetration commences if the contents are suitable.[66]

Following penetration of a cell the contents may be withdrawn immediately through the stylet by means of the pumping action of the pharynx (figure 3.8*A*, and *B*), or secretions from the pharyngeal glands may pass through the stylet into the cell where they bring about extra-corporeal digestion of the cell contents. The wholly or partially digested material is then withdrawn through the stylet by means of the sucking action of the pharynx (figure 3.8*C*). *Hexatylus viviparus*, however, feeds passively on the host cell, repeated defecation bringing about a lower pressure in the alimentary tract than in the host cell, with the result that cytoplasm flows from the host cell into the nematode.[67,68] The time taken for salivation can range from twenty seconds to two hours in different species and the time taken for complete ingestion of the cell contents can range from thirty seconds to one week. Fungal-feeding nematodes belonging to the genus *Ditylenchus* have a short stylet, a relatively small median pharyngeal bulb and use the posterior pharyngeal bulb as the main ingestion mechanism. They also have relatively short salivation and ingestion times. Parasitic *Ditylenchus* have a large median bulb, a well-developed dorsal pharyngeal gland, and carry out more extra-corporeal digestion. The median bulb is more important than the posterior bulb in ingestion and these parasitic species take a longer period of time over salivation and ingestion.[69]

Nematodes that enter the tissues of the plant and become endoparasitic first enter through the epidermis, lenticels, or stomata. When juveniles of *Heterodera* penetrate into the tissues of the host plant the stylet is withdrawn after penetration of a cell and another hole, adjacent to the first, is made in the cell wall. This is repeated several times and as each perforation is made it merges with the previous one until eventually a slit is made. This is referred to as the 'cutting cycle'. The juvenile nematode then passes through the slit in the wall of the cell and repeats the attack on a cell deeper in the plant. A juvenile of *Heterodera cruciferae* takes about five minutes to cut a slit large enough for it to pass through. When juveniles of *Heterodera* or *Meloidogyne* reach the vascular bundle within the plant they become sedentary, feeding commences, and giant cells are formed (see section 3.2.2b).[51,70] Phytophagous and mycophagous nematodes can be grouped by their methods of feeding, as described in the following sections.[51,70]

3.2.2 (a) Nematodes that bring about destructive cellular changes

Some nematodes withdraw small amounts of material from host cells and little visible damage results. For example, when *Tylenchorhynchus dubius* feeds upon root hairs a spherical mass forms at the stylet tip inside the cell. After thirty seconds the bulb pulsates and the spherical mass disappears within one minute. There are no other visible changes in the host cell and it survives the attack (figure 3.8*A*).

Most phytophagous and mycophagous nematodes affect the host cell more drastically. *Aphelenchus avenae* completely removes the cell contents of fungal hyphae in fifteen seconds (figure 3.8*B*) but *Ditylenchus myceliophagus* is a much slower feeder. There is an injection period of about forty minutes during which time fluid passes through the stylet into the hyphal cell from the dorsal pharyngeal gland; the posterior part of the pharynx then pumps for about forty minutes and withdraws the cell contents (figure 3.8*C*).

Many nematodes that feed on the cells of higher plants affect only those cells that they attack. For example, the stylet of *Rotylenchulus* penetrates the cell wall into the cell vacuole and the nematode may remain feeding on this cell for some time. The cell turns brown and dies when the nematode withdraws but adjacent cells are not affected (figure 3.8*C*). Some species of *Trichodorus* and *Hemicycliophora* secrete a feeding tube or plug, which encircles the stylet and attaches the nematode firmly to the feeding site. This enables the nematode to maintain the feeding position for long periods. *Longidorus* possesses a long stylet and this allows the nematode to feed upon the stele while remaining outside the plant. Some nematodes can alter the permeability and nature of the contents of cells adjacent to the attached cell, for example *Radopholus similis* brings about the disappearance of starch in cells adjacent to attacked cells and *Pratylenchus* affects the permeability of tissues at some distance from itself. *Ditylenchus dipsaci* brings about separation and hypertrophy of cells at some distance from itself (figure 3.8*D*).

3.2.2 (b) Nematodes that bring about adaptive cellular changes in the host

Some plant-parasitic nematodes bring about non-destructive alterations in the metabolism and structure of host cells and tissues and there are varying degrees of changes brought about by different species.

Juveniles of *Tylenchulus semipenetrans* feed on superficial cells of roots as described in section 3.2.2a. The young adult female, unlike the juvenile, penetrates more deeply into the cortex and its head becomes surrounded by six to ten altered cortical cells (nurse cells) upon which it feeds. These altered cells are normal in size but have a dense cytoplasm, an enlarged nucleus and nucleolus, and no vacuole (figure 3.8*E*). After a period of time these nurse cells disintegrate and the nematode moves further along the cortex. A development of this type of feeding is shown by *Naccobbus*, which forms galls on roots. Juveniles invade roots by intracellular migration leaving a tunnel of destroyed cells behind them. The fourth-stage juvenile and the female become sessile in the cortex where they induce the formation of a syncytical mass by breaking down the lateral walls of cells adjacent to the head of the nematode. A channel extends from the head of the nematode into the interior of the syncytical mass and the nematode feeds on this mass.

Females of *Heterodera* and *Meloidogyne* induce the formation of multinucleate syncytia or giant cells in the cortex or stele of the root (figure 3.8*F*). The giant cells produced by the plant in response to attack by *Heterodera* are apparently transfer cells and provide a localised source of protein for the nematode, which increases in weight one hundred times during development. It is suspected that roots containing these giant cells shift their metabolism in the direction of increased protein synthesis and reduce the transport of nutrients to the rest of the plant. This differentiation of unspecialised cells of the root to form transfer cells, which have a specific physiological function in the plant, is thought to be induced by the feeding demands of the nematodes and by the introduction of plant growth regulators into the incipient giant cells. The syncytia induced by *Heterodera* tend to retain remnants of cell walls but many nuclei are present in the syncytium (figure 3.8*F*).[51,70] It is worth noting that certain varieties of potato, which are resistant to *H. rostochiensis* do not form giant cells when attacked by this nematode and most juveniles develop into males. Giant cells are not essential for the development of the male. Nematodes in this group all seem to stimulate cells to increase synthesis of products upon which the nematodes feed. Similarly, at low nematode densities on tomato roots most juveniles of *Meloidogyne incognita* develop into females, but at high densities most develop into males. This is brought about by sex reversal as second-stage 'female' juveniles can develop into males that have two, instead of the normal one, testes.[18,19]

3.2.2(c) Nematodes that feed upon diatoms and blue-green algae

Many nematodes without stylets feed upon whole diatoms, blue-green algae and some other unicellular organisms, the food particles being sucked into the alimentary system by the pharynx, as in microbial feeders.

3.2.3 Carnivorous

Some stylet-bearing nematodes prey upon a range of small animals including protozoa, nematodes, rotifers, tardigrades, small oligochaetes or upon their eggs or cysts and the method of feeding is very similar to that in some phytophagous nematodes that have stylets (figure 3.8*B* and *C*). Members of the genus *Dorylaimus* are microbivorous, mycophagous or phytophagous, but some are carnivorous and feed upon other nematodes. Some nematodes are omnivorous and can feed upon a selection of different foods such as algae, protozoan cysts, fungi and nematodes. The nematode must be suitably braced against external resistances when feeding, the lips are closely applied to the prey, the stylet is thrust into it, saliva is secreted into the wound and the digested contents are then sucked out. Some predators may secrete a paralysant into the prey; however, simply puncturing the body wall of many small animals, especially nematodes, will cause reduction in body turgor pressure and this will immobolise the prey, enabling feeding to proceed without the prey struggling to escape. *Seinura tennuicaudata* halts its nematode prey as soon as the stylet pierces the cuticle, but the prey is not completely paralysed for another fifteen minutes, which suggests that secretion of a paralysant by this nematode is not essential for capture of the prey.[71] A species of *Nygolaimus*, which feeds upon small oligochaetes, has a solid stylet that is retracted after it has pierced the worm; the contents then being sucked out by the pharynx.

Many predacious nematodes do not have a stylet but possess powerful teeth or jaws (figure 3.2*E*) or small denticles in the buccal cavity and these puncture, shred, or bite the prey. Some nematodes possess both teeth and a stylet (figure 3.2*D*), the teeth holding the prey while the stylet is thrust into it (*Actinolaimus*). Many jawed predators swallow small prey whole but with larger prey they rupture the cuticle with their teeth and suck out the contents, or bite pieces off and swallow the pieces whole. *Mononchus*, for example, feeds on protozoa, nematodes, rotifers, tardigrades, and small oligochaetes. Captured nematodes are often seen whole or partially digested in the intestine of mononchids, together with hard, indigestible structures such as the stylets of nematodes or the chaetae from oligochaetes. One individual *Mononchus* killed 1332 nematodes in twelve weeks in one experiment. Predacious mononchs will sink their teeth into passing nematodes when they come into contact with the cuticle of the prey, but this response is not always elicited, which may suggest that these nematodes feed only when hungry.[72]

3.2.4 Entomophagous

Several species of nematode belonging to genera that are normally microbivorous or phytophagous, are parasitic in insects. Species of *Tylenchus* and *Aphelenchoides* live on fungi in the frass in tunnels made by bark beetles and are often transported on the bodies of these beetles. A few species, however, have become parasitic in the beetles, the juvenile nematodes living in the haemocoele of the Malphighian tubules. A species of *Delodanus* feeds on a fungus and can undergo several generations on this supply of food but if larval *Sirex* (the wood wasp) are also present then some juveniles enter the insect and develop to fertilised females by feeding on the insect. Here we have a nematode that can feed on both fungal hyphae and on insect tissues. *Neoaplectana* species are facultative parasites, which enter the intestine and invade the tissues of certain species of insect. The intestine of *Neoaplectana carcopsae* contains a bacterium, *Achromobacter nematophilus*, which is released by the nematode into the tissue and body cavity of the insect. Here both the bacterium and the nematode multiply, the nematode feeding on the bacterium and the broken-down tissues of the dying insect. When all nourishment has been exhausted the juvenile nematodes move out from the empty corpse into the soil, carrying with them the *Achromobacter* in their intestine. The cycle continues when they enter another insect.[73]

Some insect-parasitic nematodes have a reduced or non-functional alimentary tract. The juvenile stages of mermithids are parasitic in insects and the adults are free-living but neither have a functional digestive system. The juveniles possess a stylet with which they attack the viscera and fat body of the insect and the cells of the intestine of these juveniles store food materials upon which the free-living adult later exists, but nutrients are taken up across the body wall. Some nematodes that develop inside insects have only a partial covering of cuticle over the body surface, or none at all (*Bradynema*) and in these nematodes the hypodermis has become modified to form a microvillous surface and nutrients are absorbed across it. Several nematodes that attack insects may have a possible role as biological control agents; some of the potentially most useful are species of *Neoaplectana* and the mermithids, which attack mosquitoes and black flies (Simulidae).[73]

Leidynema appendiculata is an oxyurid nematode, which feeds upon the contents of the hind-gut of cockroaches. It is essentially microbivorous and uses a powerful sucking pharynx, which lacks both stylet and teeth, to ingest bacteria and particulate material present in the gut lumen of this insect.

Microfilariae (first juvenile stage) of filarial nematodes, which are transmitted by biting flies, are ingested by the insect when it feeds upon an infected host and migrate to the wing muscles of the insect where they continue their development. They are thought to feed upon the nutrients, such as amino acids, sugars and fatty acids, which were intended for the wing muscles of the insect, but they may also penetrate and feed upon the host cells by using a small buccal tooth and histolytic secretions.

3.2.5 Parasites of Vertebrates

These can be grouped by their methods of feeding:

(1) Nematodes that feed on the contents of the alimentary canal of the host (adult *Ascaris, Oxyuris, Heterakis, Ascaridia, Cosmocerca*). The lips are reduced in number in many species and the buccal cavity is usually simple. Nematodes in this group are essentially microbivorous and/or saprophagous feeders. Typical

examples are *Ascaris lumbricoides* and *A. suum* which feed on the intestinal contents of man and of the pig, and *Ascaridia galli* and *Heterakis gallinarum* (figure 3.1*B*), which feed on the intestinal and caecal contents of the domestic fowl.

Charcoal, barium and other materials have been fed to animals infected with *Ascaris* or related nematodes and these substances have been ingested by the nematodes. Incompletely digested mitochondria and other cell organelles have been found in the intestinal lumen of *Ascaris*, but it is thought that these originate from cells sloughed off the intestinal mucosa of the host. Adult *Ascaridia* take up radioactive phosphate when it is fed to the host but not when the labelled phosphate is injected in the host's blood stream showing that the nematode feeds on intestinal contents and not on the blood or tissues of the host.[48] *Cosmocerca ornata* feeds by drawing semi-digested food, bacteria, and other intestinal contents into the alimentary system by means of the powerful sucking pharynx. Histolytic secretions are produced by its pharyngeal glands and mix with the ingested bacteria as they pass into the intestine of the nematode. This nematode is able to express the fluid from the ingested food and takes mainly particulate material into the intestine. Separation of fluid from the ingested particles occurs in the pharynx when closure of the pharyngeal lumen traps particulate material in the centre of the lumen. This also forces fluid into the radial semi-circular channels and then forwards to the mouth.[53] The pharynx of *Ascaris* functions as a simple two-stage pump, in which half the lumen is filled with food at any one time (figure 3.6*B*).[54]

(2) Nematodes that feed on the mucosa of the alimentary tract or the respiratory tract of the host and have large buccal cavities, which are often armed with teeth or cutting plates (adults of *Ancylostoma, Necator, Uncinaria, Chabertia, Cyathostoma, Bunostomum, Syngamus*) (figure 3.2*F, G* and *H* and figure 3.1*A*).

These nematodes browse on the mucosa of the host's respiratory tract (*Syngamus*) or alimentary system and the hookworms of medical and veterinary importance are included in this group. Many possess a large buccal cavity, which has cutting plates or hooks at the entrance and sharp teeth at the base (figure 3.2*G* and *H* and figure 3.1*A*). The nematode feeds by drawing a plug of tissue into the buccal cavity, which it abrades with its cutting plates and teeth. Digestive enzymes and possibly an anticoagulant are secreted from the pharyngeal glands into the buccal cavity. Abraded and partially digested tissues are then pumped into the intestine by the pharynx. *Cyathostoma* draws a plug of host mucosa into the buccal cavity where buccal teeth rupture the mucosa and its underlying blood vessels releasing blood which is sucked into the animal by the pharynx. Blood passes down the pharynx in two waves, at the rate of at least one wave per second and feeding is almost continuous as freshly collected specimens always contain blood with intact erythrocytes in the anterior part of the intestine.[74] *Ancylostoma* sucks in a plug of intestinal mucosa and feeds upon detached epithetial cells and blood. Tissue drawn into the buccal cavity becomes constricted at the buccal orifice, the host tissue is broken down in the buccal cavity and is passed back to the intestine and the nematode then takes another bite of tissue. Blood is rapidly pumped into the intestine but large quantities pass out of the anus unchanged. Haemorrhage at the sites of attachment of the nematode also occurs and is important in hookworm disease.[75] Many nematodes that have well-developed buccal capsules but have no teeth or cutting plates (figure 3.2*F*) draw a plug of tissue into the capsule and carry out extra-corporeal digestion by means of pharyngeal secretions. They then ingest the fluid and semi-fluid material (*Chabertia*).

(3) Nematodes that have a small buccal cavity and feed by extra-corporeal digestion of the tissues or by puncturing cells and tissues (adults of *Haemonchus, Nematodirus, Nippostrongylus, Anisakis, Dioctophyme*).

Many nematodes that have poorly developed buccal cavities and small mouths are able to feed on tissues and blood. For example, *Haemonchus contortus* makes cutting movements at host tissues with a tooth in its mouth and feeds on the released blood, cell contents and fluids. An anti-coagulant is secreted into the wound. The adult nematode does not enter the mucosa but the fourth-stage juvenile burrows into the mucosa of the abomasum where it causes formation of a blood clot beneath which it lies and feeds.

Adult *Nippostrongylus* feed on tissues and tissue fluids of the host's intestine rather than on gut contents.[48] The ridged cuticle of the nematode abrades the mucosa when the nematode moves between the intestinal villi; histolytic enzymes that are released from the dorsal pharyngeal gland and from the sub-ventral glands associated with the excretory system attack damaged mucosal cells and bring about extra-corporeal digestion. Partially digested material is then ingested by the pumping action of the pharynx.[11,24] Sub-ventral glands are associated with the excretory system in other trichostrongyles (*Necator, Ancylostoma, Nematodirus*) and in some other groups. Enzymes have been detected in some of these glands, but the possible role of the glands in feeding and digestion is largely unexplored.[12,13] Secretions from the large dorsal pharyngeal gland of *Dioctophyme renale* produce extensive cytolysis of kidneys that it invades and the nematode feeds on the semi-digested material.

In the Ascaroidea there is a relationship between development of the pharyngeal glands and the method of feeding. Ascarids, which attack the mucosa and cause liquefaction of the tissues have a well-developed dorsal pharyngeal gland and it is probable that secretions from this gland are partly responsible for extra-corporeal digestion of the tissues (*Anisakis*). *Ascaris*, and other ascarids which do not attack the host's tissues but feed on the contents of the intestine (group 1) have a much less well-developed dorsal gland.

(4) Nematodes that possess stichocytes and feed by extra-corporeal digestion or ingestion of cell contents (*Trichuris, Capillaria*).

Adult *Trichuris*, which live partly buried in shallow, tunnel-like constructions on the surface of the gut mocusa, have a protrusible mouth stylet (not the same as those in plant-parasitic nematodes) which may be used to lacerate host cells.

Trichuris vulpis feeds on blood and *T. suis* feeds on host mucus but *Capillaria hepatica* induces the formation of specialised cells from host hepatocytes and feeds upon the cytoplasm and cell organelles of these cells. *Capillaria hepatica* also possesses a mouth stylet but it cannot be projected from the mouth in this species. The pharynx of these nematodes possesses no pharyngeal gland but large cells called stichocytes, which lie external to the pharynx and open into its lumen, are thought to produce secretions that may play a role in the digestion of food. Fluid and semi-fluid food passes back along the lumen of the pharynx in a series of waves and two muscle systems seem to bring this about. The radially arranged muscles of the pharynx dilate the lumen and a muscle sheath around the pharynx posterior to the nerve ring probably closes the lumen, thus squeezing the food along the lumen of the posterior pharynx.[52]

(5) Nematodes that live in body and tissue fluids and feed on these fluids (*Wuchereria, Dirofilaria, Onchocerca*).

Adult filarial nematodes live in lymph glands, blood vessels, coelomic spaces, and certain tissues of vertebrates. They have no food-procuring organs other than a very small mouth and narrow pharynx, which presumably take in liquid food such as serum or lymph.

The juveniles of many other nematodes pass through the blood or lymphatic systems of their hosts during migration to the respiratory tract or the alimentary system but it is not known if they feed on these fluids.

3.3 Digestive Enzymes

Digestive enzymes of nematodes appear to be related to the type of food they ingest but most nematodes are capable of digesting carbohydrates, proteins and fats to a greater or lesser extent.

3.3.1 Enzymes in the Pharynx

Most nematodes possess pharyngeal glands, which secrete substances, including enzymes, onto ingested food as it passes along the pharynx. In some species enzymes and certain other substances, such as anticoagulants or paralysing chemicals, are secreted from the pharyngeal glands and pass through the mouth or stylet onto, or into, host cells or tissues.

Many predatory nematodes carry out extra-corporeal digestion of their prey and the secretions from their pharyngeal glands probably contain several digestive enzymes. For example, *Diplogaster*, some species of which are predators, secretes proteolytic enzymes and a glycogen-splitting enzyme, and *Actinolaimus*, which possesses esterases around the base of the jaws and stylet, may use these enzymes when feeding.

There is little information about the enzymes that are present in the pharynx of free-living microbivorous nematodes. *Pontonema vulgaris* digests bacteria, which it draws in with mud from the environment. The pharyngeal glands of this nematode, which open into the buccal capsule through three teeth at the base of the capsule, show no enzymic reaction histochemically, but extracts of the pharyngeal region contain esterase and it has been suggested that one type of gland cell produces an enzyme precursor and the other type an activator, the two secretions yielding an active enzyme when mixed on discharge.[60]

More is known about enzymes and other secretions in the pharynx of nematodes that feed on fungi and higher plants. Several species secrete amylase, pectinase, cellulase, invertase, protease and chitinase in their saliva. Some of these enzymes probably assist the nematode by softening the cell walls of fungi or of higher plants and by loosening the tissues of plants. *Ditylenchus dipsaci* secretes a pectinase into the onion and brings about breakdown of the intercellular pectin, which results in the maceration of the tissues of the plant (figure 3.8*D*). It also has a cellulase, a polymethyl galacturonidase and a pectin methyl-trans-eliminase, which is activated by calcium ions. It is presumed that all of these enzymes are secreted into the tissues of the host plant and that they probably originate in the pharyngeal glands. *D. dipsaci* secretes only fourteen per cent as much amylase as *D. destructor*, which feeds on potato starch. Granules that accumulate in the ducts of the pharyngeal glands of the second-stage juvenile of *Meloidogyne javanica* prior to hatching are thought to contain enzymes that assist with penetration of the egg shell during hatching, and of the host root during feeding. Cellulase is secreted by this juvenile and probably softens the host's cell walls. Chemical and morphological changes occur in these glands after the worm has penetrated the host root and appear to be related to the changed environment of the nematode. As the parasite grows in the root the sub-ventral pharyngeal glands decrease in size and the dorsal gland enlarges. This reflects the increasing importance of the dorsal gland in the feeding activity of this nematode once it has formed giant cells in the root.[4,8,70] Activity of amylase and invertase is low throughout the gall produced by the nematode, but activity is high within the giant cells. There is no starch in the giant cells or in their immediate vicinity because hydrolysis of starch is brought about by amylase, which is either secreted by the nematode into the giant cell or formed within the giant cell following initiation by a secretion produced by the nematodes. The enzyme diffuses into neighbouring cells where it hydrolyses starch into soluble sugars that diffuse into the giant cells where they are broken down to monosaccharides. Peroxidase and a histone-like protein are also secreted by *Meloidogyne* but their function is not known.[76]

Skin- and tissue-penetrating juveniles of some animal-parasitic nematodes bring about changes in tissues of the host during entry and migration in the final host. They produce a 'spreading factor', which has a similar action to hyaluronidase, an enzyme that has been identified in only a few species. Some skin-penetrating juveniles bring about enzymic changes in the extra-cellular glycoprotein material of the skin and cause the basement membrane at the site of penetration to disappear. While some skin-penetrating juveniles seem to require enzymes to penetrate skin others seem to be able to penetrate mechanically and do not use enzymes.[8,77,78]

Proteolytic enzymes are present in the pharyngeal secretions of many nematodes and some are secreted onto the food where they bring about extra-corporeal digestion prior to ingestion by the nematode. For example, proteolytic enzymes are probably secreted into host cells by plant-parasitic nematodes. Proteolytic enzymes are present in the pharynx of adult *Ancylostoma caninum* and almost certainly play an important role in feeding and digestion in this nematode (see section 3.2.5) as the plug of mucosa that is drawn into the buccal capsule of the nematode is enzymatically broken down in less than ten minutes.[75] *Trichuris muris* burrows its pharyngeal region in the tissues of the host and has several enzymes that could be of use in carrying out extra-corporeal digestion. The anterior end of the nematode has much higher proteolytic and esteratic activity than the posterior end but so far none of the enzymes has been definitely implicated in penetration and digestion of the host's tissues.[8]

Many nematodes belonging to the Ascaroidea (*Anisakis, Contracaecum*) carry out extra-corporeal digestion and the pharyngeal glands probably secrete enzymes.

Ascaris has an amylase, a maltase, a protease, peptidases, a lipase and esterases in the pharynx and some of these enzymes will almost certainly be mixed with ingested food as it passes along the pharynx.[8]

The pharyngeal glands of *Cosmocerca ornata* contain an esterase, which is released during feeding, and bacteria are attacked as they pass down the pharynx. The esterase becomes adsorbed onto the glycocalyx of the intestinal microvilli.[53] An esterase is also produced by the pharyngeal glands of *Rhabdias bufonis* but it passes into the intestine where it acts as an endopeptidase during the breakdown of haemoglobin.[79]

3.3.2 Enzymes in the Intestine

3.3.2(a) Digestion of carbohydrates

Most nematodes are able to digest starch and glycogen and the amylases that break down these carbohydrates are similar to those of other animals, but there are minor differences in pH optima, activators and inhibitors of the enzymes. It is difficult to determine the exact origin of the enzymes of plant-parasitic nematodes because their small size precludes removal of the various regions of the alimentary tract for biochemical study. Several enzymes, including polysaccharases and glycosidases, are secreted to the outside by plant-parasitic nematodes and presumably these originate in the pharyngeal glands (section 3.3.1). Enzymes also occur in the intestine where they may be intra- or extra-cellular in function.

Amylase is present in all of the plant- and animal-parasitic nematodes that have been investigated. In plant-parasitic species the activity of the enzyme appears to be related to the nature of the food. For example, *Ditylenchus destructor* feeds on potato starch and has a much more active amylase than *D. dipsaci*, which feeds on pectin.

Enzymes that break down cellulose, pectin and chitin have been found in nematodes that feed upon fungi or on higher plants and the presence of the enzymes is apparently related to the type of feeding of these nematodes, as *Turbatrix aceti*, which feeds on bacteria and yeasts, has neither cellulase nor chitinase. Chitinase is present in *Ditylenchus myceliophagus*, which feeds upon fungi, but it is also present in *D. dipsaci* and *D. destructor*, which feed on higher plants and not on fungi. These enzymes are probably all located in thc pharyngeal glands rather than the intestine.[8,51,80]

Chitinase is present in the eggs of *Ascaris*, where it probably facilitates hatching, and in the sperm.[81,82]

'Mucopolysaccharidase' and hyaluronidase have been detected in some animal-parasitic nematodes but probably occur in the pharyngeal glands. Nematodes that feed on mucosal surfaces probably possess a 'mucopolysaccharidase'.[8]

Several enzymes that hydrolyse sugars are secreted in the saliva of some plant-parasitic nematodes but glycosidases are probably also present in the intestine of these phytophagous nematodes. Several glycosidases have been detected in the intestines of animal-parasitic nematodes, for example maltase, invertase, palatinase and trehalase are all present in the intestine of *Ascaris*, where they are localised at the brush border of the intestinal cells, but lactase is absent. Some of the glycosidases of *Trichuris* and of *Nippostrongylus* show many resemblances to the corresponding enzymes from mammalian sources and are closer to them than the glycosidases of other invertebrates. The β-acetyl aminodeoxyglucosidase, β-glucuronidase, β-galactosidase and α-mannosidase of *Nippostrongylus* are much more active than the corresponding enzymes in mammals. *Trichuris*, however, has much less β-glucoronidase and β-galactosidase activity than mammalian tissues. Both of these nematodes feed upon the mucosa of the host and, as muco-substances may be natural substrates for many glycosidases, the presence of active glycosidases in these two nematodes may be significant.[8]

3.3.2(b) Digestion of proteins

Most of the food of predatory stylet-bearing nematodes is probably digested extra-corporeally and little further digestion will take place in the intestine. Carnivorous forms (*Mononchus*) which swallow prey whole or in pieces, probably possess proteases in the intestine as digestion of prey can be observed within the living nematode. There is, however, no detailed information on these enzymes in any predatory or microbivorous nematode. Some phytophagous nematodes contain proteolytic enzymes, which are probably located in the intestine as well as in the pharynx.

Animal-parasitic nematodes that feed upon tissues of the host have higher proteolytic activity than those that feed upon partially digested food and bacteria in the lumen of the intestine. Endopeptidases attack peptide bonds in the inner regions of protein molecules and split them into shorter polypeptide chains. No pepsin-like enzyme has been detected in nematodes but endopeptidases with pH optima of 6.0–6.2 are present in the intestine of *Ascaris, Strongylus* and *Leidynema.*[8] In some respects these endopeptidases are similar to cathepsins, but activators of mammalian cathepsins give negative or doubtful results. Several animal-parasitic nematodes (*Ascaris, Rhabdias*) break down ingested haemoglobin and must have active endopeptidases in the intestine.[8,83,84] *Rhabdias bufonis* has an esterase in the intestine where it acts as an endopeptidase.[79] The free-living nematode *Pontonema vulgaris* digests haemoglobin in the intestine but this is brought about by esterases that are produced by the pharyngeal glands. Carbonic anhydrase is produced by the intestinal cells of this nematode and may be associated with the production of hydrochloric acid.[60]

Exopeptidases attack the terminal peptide bonds of polypeptide chains, which have been produced by the action of endopeptidases on proteins, to liberate amino acids. The intestine of *Ascaris* contains leucine aminopeptidase, a tripeptidase, alanylglycine peptidase and glycylglycine peptidase. One of the intestinal aminopeptidases of *Ascaris* has been purified and shown to be different from two similar enzymes found in uterine extracts of the same nematode.[8] These enzymes have not been studied in free-living or plant-parasitic nematodes.

3.3.2(c) Digestion of fats

Lipases hydrolyse fats to give fatty acids and glycerol while esterases split organic and inorganic esters.

Lipases are probably present in the intestine of most nematodes because oils, fats and lipid complexes with carbohydrates and protein are found in most of their foods. However, lipase has not been detected in *Graphidium* nor in *Trichuris.* The lipase of *Ascaris* intestine is activated by the bile salt sodium glycocholate and inhibited by sodium taurocholate but that of *Strongylus edentatus* is inhibited by both substances. *Strongylus* feeds on the tissues of the host and has a lipase that is twelve times more active than the lipase of *Ascaris,* which feeds on the gut contents of the host.[8,48] Lipase has been detected in intestinal cells of *Rhabdias bufonis.*[79]

Non-specific esterases are present in the intestine of most nematodes. Those in the intestines of *Ascaris* and *Trichuris* are similar to some mammalian esterases.[8]

3.3.3 Enzymes in the Sub-ventral (Excretory) Glands

The Strongylidae possess a pair of large glands associated with the excretory system (see sections 1.2.8 and 5.3). These glands are apparently secretory rather than excretory in function and one of their functions in some nematodes (*Nippostrongylus, Stephanurus*) is to assist in extra-corporeal digestion.[11,24,85] They contain a number of enzymes, including leucine aminopeptidase, non-specific esterases, proteases and, surprisingly, acetylcholinesterase.[11] Not all of these enzymes are present in all species; adults of *Nippostrongylus brasiliensis, Oesophagostomum radiatum* and *Trichostrongylus colubriformis* contain large amounts of acetylcholinesterase but there is little or none of this enzyme in these glands in other species, such as *Oesophagostomum venulosum*.[12,13] Some of these enzymes are secreted through the excretory pore and probably assist in the digestion of host tissues around the anterior end of the nematode. Acetylcholinesterase is also secreted onto the host tissues but its exact role is unknown. It may alter the permeability of host cells or tissues and allow serum to flow into the intestinal lumen where it will be available for ingestion by the nematodes. On the other hand it may affect sensory receptors or the muscles of the intestine and prevent local spasm of the intestine when the nematode attacks the mucosa.[11,24]

3.4 Secretion in the Intestine

Secretory activity in the intestine has been described in relatively few nematodes. The intestinal wall of *Pontonema vulgaris* contains both glandular cells and columnar absorption cells. Columnar cells develop into the glandular type, which then carry out merocrine secretion before being extruded into the lumen of the intestine.[60] Apocrine secretion occurs in the intestine of *Ascaris* and of *Ascaridia galli*, parts of the cell containing secretory material becoming detached from the intestine and disintegrating in the lumen.[86] Merocrine secretion, in which droplets coalesce and pass through the cell border into the lumen, also occurs in *Ascaris*. The anterior part of the intestine of *Ascaris* contains many more secretory cells than the mid-intestine and posterior intestine.[8]

3.5 Uptake of Nutrients

Most uptake of nutrients by nematodes takes place through the intestine but a few exceptions do occur. For example, the alimentary system of juvenile *Mermis nigrescens* is atypical in that it consists of a non-muscular pharynx and an intestine that functions as a storage organ. These juveniles take up amino acids and glucose through their cuticle, rather than through their gut, and absorption is mediated by an active transport system. Trehalose, which is the major carbohydrate in the blood of their insect host, is not taken up by these juveniles. Glucose is taken up faster by fourteen-day-old than by twenty-one-day-old juveniles; this may be related to the thinner cuticle or to the greater needs of the younger worms.[87,88] *Sphaerularia bombi* is even more unusual in that it takes up nutrients across the everted and enlarged uterine wall. Areas of the body wall of adult *Bradynema*, which lives in the haemocode of insects, have no cuticle and the exposed hypodermis is extended into microvilli. It is probable that nutrients enter the nematode across this modified body surface.[28] The cuticle of *Ascaris* is not permeable to glucose and this is probably the case in most nematodes that have a functional alimentary system.[8]

Fluid moves freely in either direction across the intestinal wall of *Ascaris* in response to osmotic gradients, but under isosmotic conditions fluid always moves from the mucosal to the psuedocoelomic side of the intestine. Movement of fluid from the mucosal to the pseudocoelomic side is greatly increased when oxygen is available. Glucose increases fluid movement and is also transported across the wall of the intestine against a concentration difference. Fluids appear to enter the cell from the lumen, pass into the intercellular spaces, which dilate during water transport and then pass into the body cavity (figure 3.9). Movement of glucose across the intestine is dependent upon the presence of sodium ions in the lumen of the intestine.[89]

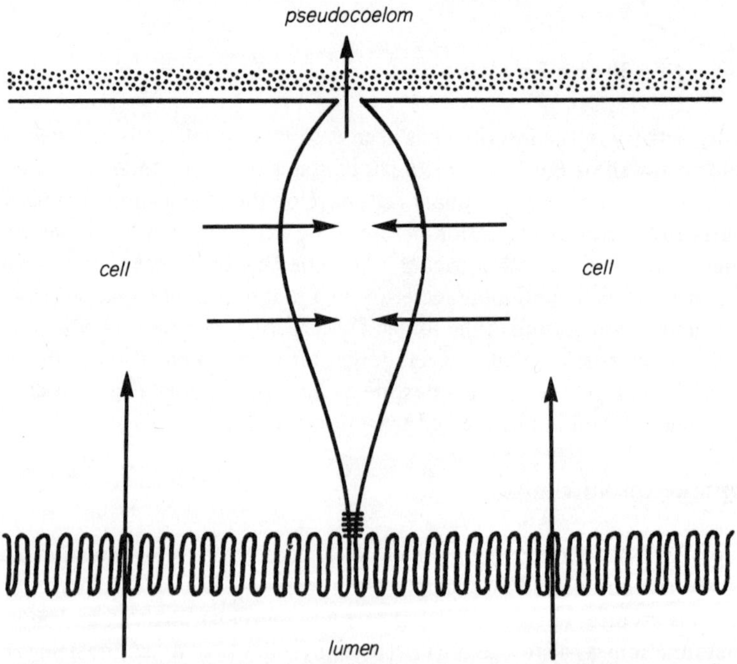

Figure 3.9 The flow of fluid (shown by arrows) from the intestinal lumen into the intercellular spaces between the intestinal cells and then into the pseudocoelom of *Ascaris* when conditions are isosmotic. [Based on electron micrographs and a description by Harpur and Popkin[89] and on Schmidt-Nielson, *Fed. Proc.*, 30 (1971).]

Both 3-O-methylglucose and fructose move across the wall of the intestine from the intestinal lumen into the pseudocoelomic fluid if glucose is present and if the gas phase comprises 95 per cent nitrogen and 5 per cent carbon dioxide, but this movement is markedly reduced in the presence of high partial pressures of oxygen. The rate of movement across the intestine is greatest in the posterior intestine.[90] Galactose is not transported across the wall of the intestine under these conditions. The rate of movement of fatty acids, which also move from the lumen of the intestine into the pseudocoelomic fluid of *Ascaris*, is a function of the concentration of the fatty acid present in the intestinal lumen of the nematode and is facilitated when the fatty acid is complexed with a bile salt and when glucose is present in the medium. The fatty acids move into the intestinal cells by means of a passive process and energy from the catabolism of carbohydrate is required to move the fatty acids from the cells into the pseudocoelomic fluid. Sterols are not actively transported across the intestine.[91]

The contents of the intestine move about as a result of the locomotory activities of nematodes and the ingestion of more food, which gradually pushes ingested food along the intestine. In a few species (*Aspiculuris, Leidynema*) a muscle network around the intestine brings about peristaltic movements, which help to move the contents of the intestine backwards and forwards.

3.6 Nutrient Requirements and Growth *in vitro*

Several free-living, plant-parasitic and animal-parasitic nematodes have been grown axenically and *in vitro* and studies of these are now leading to an understanding of the nutrient requirements of these nematodes. Much of the early work was carried out on free-living, microbivorous nematodes such as *Caenorhabditis* and *Turbatrix*, or on the insect-parasitic/saprophagous nematode *Neoaplectana.* The culture media for most of these nematodes consists of a defined basal medium – which usually contains amino acids, carbohydrates, sterols, vitamins and essential salts – to which is added a tissue supplement such as liver extract or chick embryo extract.[92] Using this type of medium it has been possible to determine which vitamins and amino acids are essential for growth and reproduction. Many animal-parasitic nematodes can now be grown axenically in culture but the media contain one or more organic materials that are chemically undefined. Nonetheless, there has been a rapid increase in knowledge of the nutritional requirements of these parasites (and their free-living stages) in recent years and attempts to develop a fully defined medium are now concentrating on elucidating the essential components of the tissue supplements. For example, it is now known that the tissue supplements supply sterols and phorphyrin or haem compounds to the nematodes in culture. It is also thought that the tissue supplements provide a suitable particle size, which is easily ingested, which stimulates feeding, and which carries the sterols and haem growth factors. The gaseous phase of the culture medium is very important, especially in the culture of animal-parasitic nematodes when the gaseous mixture usually contains 5 per cent carbon dioxide, little or no oxygen and 25 per cent nitrogen, but some nematodes require up to 10 per cent oxygen as well as 5 per cent carbon dioxide in nitrogen.

The culture of nematodes is a diverse and rapidly moving field of research and the reader is advised to read the excellent reviews on this subject as well as studying the relevant journals of nematology and parasitology.[92,93]

3.7 Defecation

Little work has been done on defecation in nematodes. The internal pressure of the intestinal lumen rises as the pharynx pumps food into it. This results in an increase in volume, which lengthens the nematode and stretches the posterior end, bringing about a stretch reflex, which causes the anal dilator muscles to contract. Defecation commences and brings about shortening of the body. This in turn causes the tail muscles to contract; the body of the nematode shortens rapidly and a jet of faeces is forced out of the anus. Defecation takes place at regular intervals when some individuals feed continuously for long periods, suggesting that defecation occurs when a fixed volume or length is reached.[9,63]

A sphincter muscle, which encircles the intestinal–rectal junction of many nematodes, keeps the intestinal–rectal junction closed and prevents movement of food from the lumen of the intestine into the rectum or cloaca. Periodic opening of the sphincter allows faeces to pass from the posterior end of the intestine into the rectum and out of the anus when the anal dilator muscles contract. There is a network of muscle around the posterior end of the intestine of a few nematodes (*Heterakis*) and at defecation this muscle contracts, the sphincter muscle relaxes (or is forced open by the increased pressure in the posterior intestine), the anal dilator muscles contract and the faeces shoot out of the anus.[10]

A few nematodes (*Aphelenchoides blastophthorus*) contain a pre-rectal valve, which closes at defecation and separates the main part of the intestine from the pre-rectum. An intestinal–rectal valve separates the pre-rectum from the rectum and opens during defecation. During defecation, which occurs shortly after feeding has stopped, the faeces are first isolated in the pre-rectal region of the intestine and in the rectum by closure of the pre-rectal valve and dilation of the rectal valve. They are expelled when the anus opens and the pre-rectum and rectum collapse.[63]

3.8 Anti-enzymes and Anti-coagulants

Nematodes that inhabit the alimentary tract of their hosts are not digested by the host unless they are dead or the cuticle is damaged. It is probable that the living organism possesses properties that are similar to those of the epithelium of the alimentary tract of the host in this respect and antienzymes may not be necessary. However substances that inhibit digestive enzymes of the host have been found in a few nematodes, for example *Ascaris* contains pepsin, cathepsin E, trypsin and chymotrypsin inhibitors.[8,94] It is suspected that anti-enzymes produced by some animal-parasitic nematodes may reduce the level of protein digestion in the host.

The amphidial glands of *Ancylostoma caninum* and of *Syngamus trachea* contain a substance that inhibits coagulation of host blood and is thought to be secreted by the nematodes during feeding. This substance in *Ancylostoma* prolongs prothrombin time and also interferes with collagen- or ADP-induced platelet aggregation. Anticoagulants are thought to occur in a few other sanguivorous nematodes.[8]

4 Metabolism

4.1 Carbohydrate Metabolism

4.1.1 Distribution of Carbohydrates

Glycogen is stored mainly in the hypodermis, the non-contractile part of the muscles, the intestine and the epithelial cells of the reproductive organs of nematodes. Free-living and plant-parasitic nematodes have been little studied, but they apparently store less glycogen than animal-parasitic species (table 4.1). This is probably related to their predominantly aerobic existence, which means that they are able to use stored lipids as a source of energy. A fungal-feeding species of *Aphelenchoides* has 2–9 per cent of its dry weight as polysaccharide and 85 per cent of this is glycogen. *Aphelenchus avenae* has 8 per cent and *Caenorhabditis* sp. 3.3 per cent of their dry body-weight as glycogen.[95,96]

The amount of glycogen in the tissues of animal-parasitic nematodes varies from species to species and also from stage to stage within species, but it is usually present in considerable amounts and provides the major reserve store of energy (table 4.1). In *Ascaris* the amount of glycogen present is about 20 per cent of the dry weight of the animal and about 70 per cent of the dry weight of the muscle is glycogen. However, the free-living juveniles of many animal-parasitic nematodes contain very little glycogen, the energy reserve being mainly lipid; for example, the free-living juveniles of human hookworms and of *Nippostrongylus* contain little, if any, glycogen, and none can be detected in microfilariae of *Acanthocheilonema perstans.* Glycogen does occur in some juvenile nematodes however; the juveniles of *Trichinella spiralis* and of *Eustrongylides ignotus* containing significant amounts of it (table 4.1).[8] In unembryonated eggs of *Ascaris*, glycogen accounts for 7.8 per cent of their dry weight and in embryonated eggs it makes up 6.9 per cent of their dry weight. This lower percentage of the dry weight, compared with the 20 per cent of the adult worm, is probably correlated with the aerobic development of these eggs and their utilisation of stored lipid (see section 4.2.2).[8,97]

Glycogen disappears rapidly from the tissues of nematodes when they are not fed. In *Ascaridia galli* there is a rapid decline in the amount of glycogen present in the tissues of the nematode when the host (the fowl) is not fed and after three days the nematode is expelled from the host. There is also a marked reduction in the amount of glycogen in adult female *Thelastoma*, which are parasitic in the hind-gut of cockroaches, when they are starved *in vitro. Ascaris* consumes 1.3 g of glycogen per 100 g body weight in twenty-four hours when starved under anaerobic conditions and 0.8 g under aerobic conditions.[8] Encysted muscle-stage juveniles of *Trichinella spiralis* can survive anaerobiosis for several days by fermenting endogenous carbohydrates, but the intestinal stages do not survive more than twelve hours without exogenous glucose.[98] *Aphelenchus avenae* uses its stored glycogen when it is starved under anaerobic conditions, but uses stored lipid when starved under aerobic conditions. *Caenorhabditis*, however, uses its glycogen reserves when starved under aerobic or anaerobic conditions.[99] *Panagrellus* and *Turbatrix* use their glycogen when starved under aerobic conditions but some lipid is also used.[100] The juveniles of *Ascaris*, and of several other species that migrate through the body of the host before settling at their final location, increase their glycogen stores during the migration. This may be related to a change in their physiology associated with a decrease of oxygen in the environment and therefore a greater reliance on glycogen rather than on lipid as an energy reserve. Differences exist in the utilisation of glycogen by embryonated eggs of different animal-parasitic species. For example embryonated eggs of *Ascaris*, when kept at room temperature for up to two years, show no demonstrable diminution in their glycogen reserves, whereas there is rapid depletion of polysaccharide in the juveniles of *Enterobius vermiculuris* if the eggs are incubated *in vitro.*[8]

Table 4.1 Glycogen content of several nematodes[8,92]

Species	Glycogen in per cent of Dry Weight	Habitat
Ascaridia galli	19.8	Intestine
Ascaris lumbricoides	14–24	Intestine
Nippostrongylus brasiliensis	3.3	Intestine
Heterakis gallinarum	11.8	Caecum
Dirofilaria immitis	10	Heart
Eustrongylides ignotus (juvenile)	28	Various organs
Litomosoides carinii	5	Pleural cavity
Porrocaecum decipiens (juvenile)	55	Muscle
Trichinella spiralis (juvenile)	15.9	Muscle
Aphelenchus avenae	8	Soil
Caenorhabditis sp.	3.3	Soil
Ascaris lumbricoides (embryonated eggs)	6.9	Soil
Panagrellus redivivus	6.5	Sour paste
Turbatrix aceti	19	Vinegar

The only glycogen that has been studied in detail in nematodes is that of *Ascaris.* Its heat of combustion is 4212 or 4125 cal g^{-1} (slightly different results were given by two different authors) and it has iodine-staining properties which differ from those of vertebrate glycogens. The molar mass of the ascarid glycogen is much greater than that of mammalian glycogen and it is divided almost equally into a variety with a molar mass of 450 million and one with a molar mass of 50 million. The average chain length is eleven to thirteen.[8]

Sugars are also present in the tissues and in the body fluids of nematodes; for example glucose and fructose are present in *Ditylenchus dipsaci* and *Aphlenchoides ritzemabosi*, and trehalose and mannose have been detected in a fungus-feeding species of *Aphelenchoides.*[96] The intestine and the body wall of *Ascaris* contain more glucose than trehalose, but the disaccharide is present in greater amounts in the muscles, the reproductive organs and the pseudocoelomic fluid. Trehalose is apparently the most abundant low-molar-mass carbohydrate in nematodes. Glucose and trehalose, which comprise 5–7 per cent of the dry weight, form reserves that are used extensively when *Nippostrongylus brasiliensis* is starved in saline.[8,109]

Unusual glycosides, known as ascarosides, occur in the egg shell of *Ascaris lumbricoides* and *Parascaris equorum.* Free and esterified ascarosides have also been detected in *Ascaris columnaris, Ascaridia galli* and *Toxocara cati* (all ascarids), in *Passalurus* sp. and almost certainly in *Leidynema* sp. (oxyurids), and in a free-living rhabditid, but they were not detected in *Turbatrix aceti* nor in *Panagrellus redivivus.*[101,102]

These ascarosides have been identified as glycosides in which the glycone (sugar) component is ascarylose (3,6-dideoxy-L-arabinohexose); the aglycone is one of a series of secondary monols and diols containing twenty-two to thirty-seven carbon atoms. These high-molar-mass glycosides are apparently unique to nematodes. Ascarylose is synthesised from glucose in the ovary of *Ascaris.* The ascarosides (75 per cent), together with protein (25 per cent), form the innermost layer of the egg shell.[8,103]

Chitin is found in nematodes, but only in the egg shell; it is synthesised in the egg and not in the oogonia, the oviduct or the uterus. During egg shell formation in *Ascaris*, acetate is liberated from ascaroside esters at the same time that chitin synthesis is occurring and it may provide the acetyl residues of the newly synthesised N-acetylglucosamine.[8,102,103]

4.1.2 Intermediary Metabolism

In most animal cells carbohydrate is broken down by a combination of glycolysis (the Embden–Meyerhof pathway) and oxidative decarboxylation (the tricarboxylic acid, TCA or Krebs' cycle). The Embden–Meyerhof pathway converts glucose or glycogen to pyruvate (figure 4.1). The resultant pyruvic acid may then be reduced to lactic acid or may produce acetyl coenzyme A (figure 4.2), which then undergoes controlled oxidation in the TCA cycle. When isocitrate, α-ketoglutarate, succinate and malate are oxidised in the TCA cycle they each lose two electrons and two protons. The electrons are accepted by a series of electron carriers on the cristae of the mitochondria and pass along the chain of carriers until they reach oxygen. Mitochondria use the energy of oxidation to form ATP from ADP and inorganic phosphate; that is, they undergo oxidative phosphorylation.

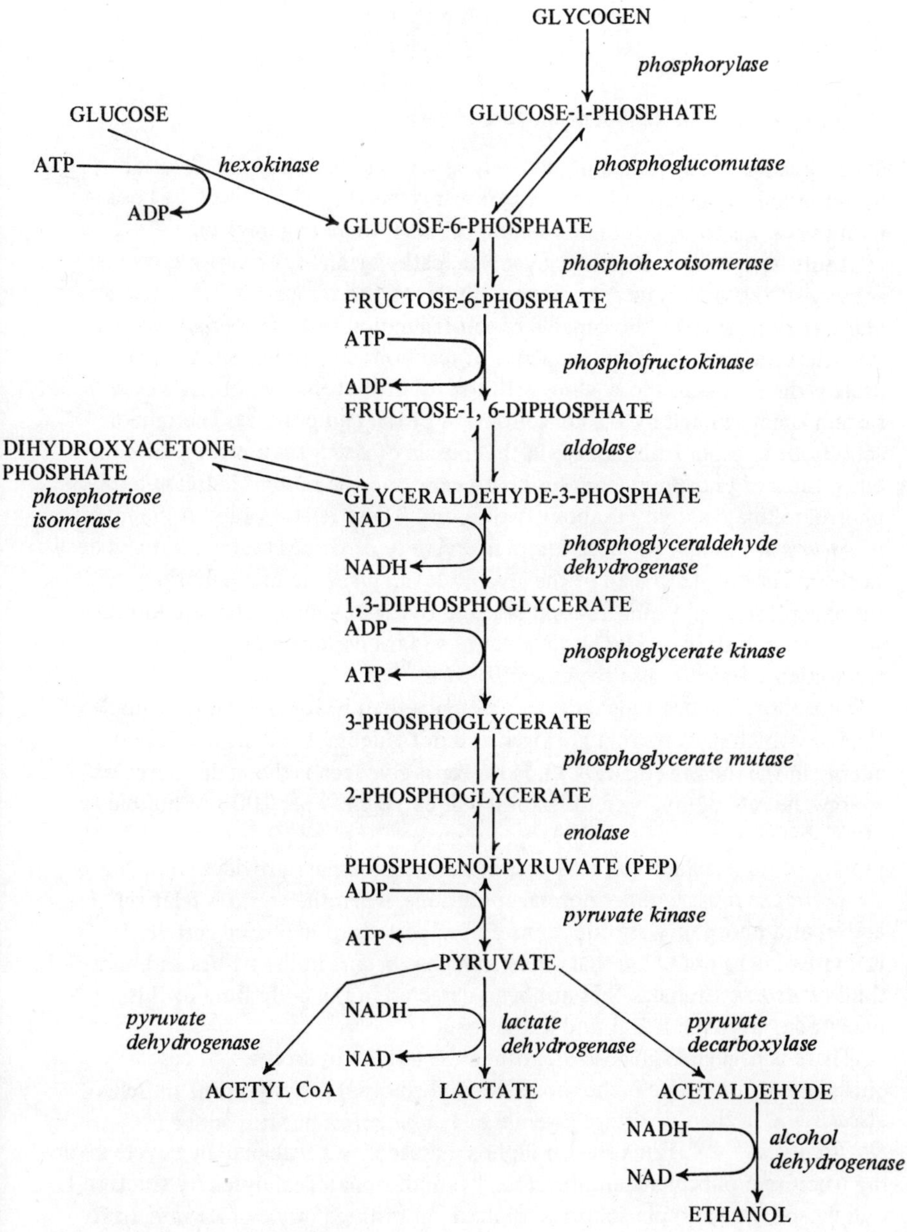

Figure 4.1 Embden–Meyerhof pathway

Aerobic metabolism of one mole of glucose can yield thirty-six moles of ATP, whereas anaerobic glycolysis yields only two.

4.1.3 Glycolysis: the Embden–Meyerhof Pathway

Glycolysis has been studied in relatively few nematodes, and most work has been limited to detection of the enzymes involved in the pathway, but this pathway seems to be a common feature of nematodes (figure 4.1).[8,104,105,106]

Adult *Ascaris* have an active glycolytic pathway and the relative rates of activity of the glycolytic enzymes are similar to those found in other tissues. Many enzyme systems are capable of self-regulation of their over-all reaction rate; the end product of the sequence of reactions can inhibit an enzyme – usually the first – in the system, with the result that the rate of the whole system is determined by the concentration of the end product. The rate of activity of phosphofructokinase in the muscle of *Ascaris* is equal to the combined rates of phosphorylase plus hexokinase and the ratio of aldolase to phosphofructokinase activity is about two to one. These relationships in the muscle of *Ascaris* are similar to those found in a variety of tissues that exhibit glycolysis. In the muscle of *Ascaris* all of the glycolytic enzymes are in equilibrium, with the exception of phosphofructokinase and pyruvate kinase. The rate-limiting step in glycolysis in all glycolytic systems so far investigated is phosphofructokinase and it is also the case in *Ascaris.*[104]

Phosphorylase has a higher rate of activity than hexokinase in the muscle of *Ascaris*, which suggests that glycogen and not glucose is the main source of energy in the muscle (figure 4.1). However if glycogen is the main source of energy the rate of glycogen utilisation will be 10 g h^{-1} per 100 g of muscle at 30 °C, which means that the glycogen reserves (13 g per 100 g of muscle) will rapidly become exhausted. It is possible that hexokinase provides enough substrate for the muscle under normal conditions, when the worm is relatively inactive, and phosphorylase functions during periods of increased activity.[104] It is interesting to note here that while trehalose occurs in the tissues and body fluid of *Ascaris*, trehalase has not been detected in the body fluid of this nematode, although it is found in the tissues.[8]

There is negligible glucose-6-phosphatase (which hydrolyses glucose-6-phosphate to inorganic orthophosphate and glucose) activity in the muscle of *Ascaris* and in the third-stage juvenile of *Haemonchus* but it is active in *Obeliscoides.*[104,106] Fructose-1,6-diphosphatase may participate in a cycle involving fructose-6-phosphate and fructose-1,6-diphosphate, catalysed by fructose-1, 6-diphosphatase and phosphofructokinase, in resting muscle of *Ascaris*, or it could be concerned with gluconeogenesis (section 4.1.14). There are significant amounts of glycerol-1-phosphate dehydrogenase activity and no alcohol dehydrogenase activity in the muscle of *Ascaris* and, therefore, an α-glycerophosphate shunt cannot occur in this nematode.[104]

4.1.4 Pasteur Effect

Under anaerobic conditions an increased rate of glycolysis, which results in greater use of substrate, is known as a Pasteur effect. It is a mechanism whereby cells are able to compensate for the lower yield of ATP that occurs when the oxygen supply is restricted. Nematodes in which a Pasteur effect has been detected are adult *Nippostrongylus brasiliensis*, adult *Litomosoides carinii*, *Dirofilaria insignia*, *D. uniformis*, third-stage juveniles of *Ascaris* and third-stage juveniles of *Haemonchus contortus.*[107,108]

4.1.5 Production of Lactic Acid and the Oxygen Debt

When vertebrate muscle is exercised under anaerobic conditions it uses pyruvate as an electron acceptor and thus NADH, which is formed during glycolysis, reverts to NAD. Lactate dehydrogenase is the enzyme that brings about this reduction of pyruvate and results in the production of lactic acid. Thus the formation of lactic acid gives the muscle a supply of ATP derived from glycolysis when oxygen is in short supply. Lactate is the final product of this reaction, not because muscle needs the substance but because it needs to utilise pyruvate as an electron acceptor. In vertebrate muscle, however, fatigue sets in when lactate accumulates at the site of the formation and this slows down the reaction.

Some nematodes form large quantities of lactate under anaerobic conditions (*Litomosoides*, *Brugia* and *Dirofilaria*) and excrete the lactate formed;[105] these species presumably escape muscle fatigue. In vertebrates the lactate is subsequently metabolised when oxygen becomes available and there is then higher-than-normal oxygen consumption while the lactate is metabolised. This is called repayment of an oxygen debt. In those nematodes that excrete the lactate produced there is little or no post-anaerobic increase in oxygen consumption after a period of anaerobiosis because the carbohydrate consumed under anaerobic conditions is equal to the amount of acid excreted, as in *Litomosoides*, or there is only a small post-anaerobic rise in oxygen consumption, as in *Ascaris* and the juveniles of *Eustrongylides.*[8,109]

Lactate dehydrogenase has relatively low activity in some nematodes and these produce little lactate under anaerobic conditions. *Ascaris* produces little lactate although it has a reasonably active lactate dehydrogenase in the cytoplasm (cytosol) of its muscles but this may be correlated with the fact that most pyruvate is produced in the mitochondria in this nematode (section 4.1.8).[104,105,108]

Strongyloides ratti is an interesting nematode as it can undergo several complete, free-living generations between the first-stage juvenile produced by the parasitic female and the infective-stage juvenile that is eventually produced by free-living females under certain conditions, and which invades the rat. These infective-stage juveniles develop to adult females, which produce eggs in the intestine of the rat. Some of the eggs hatch in the alimentary tract and eggs plus juveniles are passed in the faeces where they develop into free-living generations again. Pyruvate kinase and lactate dehydrogenase are present in all of the free-living stages of this nematode but they are significantly reduced in the parasitic females (table 4.2). This is almost certainly related to the different metabolic pathways in the free-living and the parasitic stages (section 4.1.8).[110]

Both *Aphelenchus avenae* and *Caenorhabditis* exhibit a small, post-anaerobic oxygen debt, but only for up to sixteen hours anaerobiosis and only with fresh, unstarved individuals. This can be correlated with the fact that both species produce lactic acid as the principal end-product of glycolysis during the first twelve to sixteen hours, after which time there is a shift to ethanol production.[95]

The free-living nematodes *Enoplus brevis* and *E. communis* do not show a post-anaerobic rise in oxygen consumption[111,112] and may, if they possess this pathway, excrete the acid produced under anaerobic conditions. This assumes, however, that these nematodes produce lactate, are active, and have a high metabolic rate under these conditions; but both *E. brevis* and *E. communis* are immobile in the absence of oxygen, as are many species of nematode including some animal-parasitic species (*Nippostrongylus brasiliensis*), some mycophagous species (*Aphelenchus avenae*) as well as some other free-living species (*Caenorhabditis briggsae*).[92,111,112,113] It is possible that in many of these species the metabolic rate is extremely low under anaerobic conditions and that, after an initial period of acclimation, neither lipid nor glycogen stores are used, as in *Aphelenchus avenae*[99] and therefore little or no acid is produced.

4.1.6 Oxidative Decarboxylation: the Tricarboxylic Acid Cycle

Most of the NADH, which is essential for the process of oxidative phosphorylation is produced by a series of reactions known as the tricarboxylic acid (TCA) or Krebs' cycle (figure 4.2). As well as its importance in the production of energy, the cycle is also important in the interconversion of carbohydrates, fats and certain amino acids.

The TCA cycle is a catalytic operation which brings about the total combustion of the acetyl groups of acetyl co-enzyme A (figure 4.2). The reactions form a cycle: acetyl co-enzyme A is fed in at the beginning of the cycle and, as the reaction completes one cycle, two carbon atoms are eliminated as carbon dioxide and reducing power – in the form of three moles of NADH and one of $FADH_2$ – is produced. The production of this reducing power, which is used in oxidative phosphorylation to produce ATP (figure 4.4), is one of the main functions of the TCA cycle.

4.1.7 The TCA Cycle in Nematodes

Some nematodes possess all of the enzymes of the TCA cycle, but others apparently possess only some of the enzymes. Evidence for the existence of a complete cycle in nematodes must be treated with some caution as in most cases the evidence relies upon the oxidation of intermediates of the TCA cycle by homogenates of the nematodes or the histochemical demonstration of some of the enzymes in the tissues. This evidence must be supported by evidence that the cycle operates in the intact nematode and, moreover, under the gaseous conditions in which the nematode normally lives.

As would be expected, most nematodes that live under aerobic conditions have a functional TCA cycle, but a number of animal-parasitic nematodes that live in low oxygen tensions also have a TCA cycle. The free-living nematodes *Panagrellus redivivus*, *P. silusiae*, *Turbatrix aceti*, *Ditylenchus triformis*, and the free-living generations of *Strongyloides ratti* appear to have a functional TCA cycle.[8,100,110,114]

The occurrence of most of the TCA cycle has also been demonstrated in the infective juveniles of *Trichinella spiralis*, in adult *Nematodirus*, *Neoaplectana* and *Ascaridia.*[8] *Dirofilaria immitis* has TCA-cycle enzymes but it has low levels of aconitase and isocitrate dehydrogenase; and malate dehydrogenase activity can only be demonstrated in the direction of malate formation, which suggests that it may operate a pathway similar to that described for *Ascaris* rather than the TCA cycle.[105,115] The presence of the glyoxalate cycle (figure 4.7 and section 4.2.2) and the conversion of acetate to glycerol in *Turbatrix aceti*, *Panagrellus redivivus* and *Caenorhabditis briggsae* suggests that at least part of an operative TCA cycle occurs in these three species of free-living nematode.[100,116,117] A number of free-living and plant-parasitic nematodes (*Enoplus brevis*, *E. communis*, *Caenorhabditis* sp., *Meloidogyne javanica* and *Aphelenchoides avenae*) are active in the presence of oxygen but become quiescent when oxygen is absent and female *Nippostrongylus* require an oxygen tension greater than 12 mm Hg to maintain their motility,[113] which suggests that they have an aerobic metabolism. Many nematodes survive anaerobic conditions for varying periods of time but they are inactive and their metabolic rate is probably very low. Survival of nematodes under anaerobic conditions does not therefore necessarily infer that they normally use anaerobic pathways as the sole source of energy.

It appears that the TCA cycle functions in some free-living nematodes, in some plant-parasitic and in some animal-parasitic nematodes.

4.1.8 Alternative Pathways of Carbohydrate Metabolism in Nematodes

The TCA cycle appears not to function, or is insignificant in the production of energy, in several species of nematode.[105,108] The reduction of fumarate to succinate appears to be an important pathway for the elimination of hydrogen under anaerobic conditions in these nematodes (figure 4.3). In the muscle of *Ascaris*, which does not possess a complete TCA cycle, glycogen or glucose is broken down by glycolysis, as previously described, as far as phosphoenolpyruvate (PEP) (figure 4.1). This does not then form pyruvate as in the classical glycolysis sequence, but is carboxylated through the action of phosphoenolpyruvate carboxykinase (PEPCK) and the fixation of carbon dioxide into PEP, to form oxaloacetate. This is brought about because there is a deficiency of

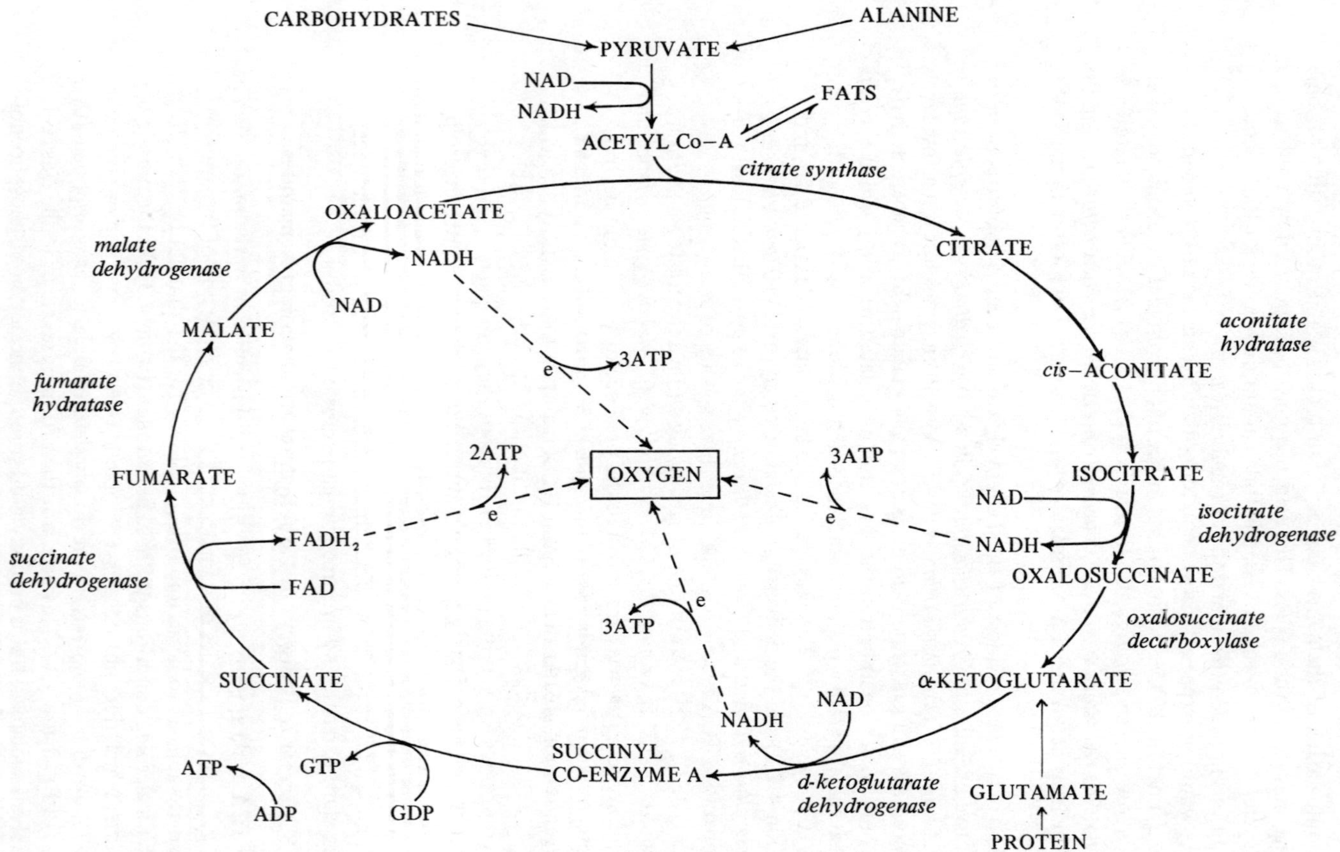

Figure 4.2 Tricarboxylic acid cycle. (--e→ represents the flow of electrons along the electron transport chain).

pyruvate kinase (PK) and high activity of PEPCK (table 4.2). Oxaloacetate is reduced to malate by a cytoplasmic malate dehydrogenase and the malate passes into the mitochondria where it becomes the mitochondrial substrate. The NADH that is required for the reduction is made available during the formation of PEP from glycogen (figure 4.3). Malate translocase may be involved in this movement of malate into the mitochondria. Malate then undergoes a dismutation reaction within the mitochondria.[108] NADH is obtained by the oxidative decarboxylation of some of the malate to pyruvate and carbon dioxide, thus giving rise to pyruvate in the absence of PK. It is important to note that this pyruvate is formed within the mitochondrion. Pyruvate is usually produced by the Embden–Meyerhof pathway within the cytoplasm of the cell of most organisms.

The reaction is catalysed by a mitochondrial, NAD-linked malic enzyme (figure 4.3). The NADH formed by this reaction is then used to reduce malate to succinate via fumarate through the action of fumarate reductase (succinate dehydrogenase functions as a fumarate reductase in the muscle of *Ascaris*). This reduction of fumarate to succinate is accompanied by the phosphorylation of ADP to form ATP. Pyruvate formed within the mitochondrion may serve as a precursor for acetate within the mitochondrion. This reaction may result in additional reducing power, which may enter into the reductive formation of the volatile fatty acids, which are such characteristic end products of carbohydrate metabolism in adult *Ascaris* (figure 4.3). This pyruvate is not normally reduced to lactate, possibly because it may not readily diffuse out of the mitochondrion into the cytoplasm and lactate dehydrogenase is present only in the cytoplasm. This could explain why *Ascaris*, although it has a functional lactate dehydrogenase, excretes only small amounts of lactic acid. This system can operate if the tissue is homogenised but it probably does not do so in the intact animal. It should be noted however that there appear to be two types of iso-enzyme of lactate dehydrogenase in the tissues of other animals; one is found in tissues where rapid glycolysis occurs and has a high affinity for pyruvate as electron acceptor, the other is found in tissues that actively respire and has a low affinity for pyruvate.

The nature of the end product formed in several animal-parasitic nematodes (and in some other parasitic worms) apparently depends on the competition for PEP by PEPCK and PK and could account for the different proportions of succinate and lactate that are produced by certain nematodes. Ratios of the activities of PK and PEPCK give some indication of the nature of the pathway (PEP to pyruvate if PK is most active or PEP to oxaloacetate if PEPCK is most active) and hence the predominant end product of glucose metabolism (lactate if PK is most active and succinate if PEPCK is most active) (table 4.2). The ratio in *Ascaris* muscle is 0.04, which explains why large amounts of succinate and its derivatives are produced by this nematode; that for adult *Nippostrongylus* is 1.86, which suggests a greater emphasis on lactate production but with some succinate also being produced, whereas in *Litomosoides carinii* the ratio is 2.8 and the emphasis is on lactate and acetate production (table 4.2). It is possible,

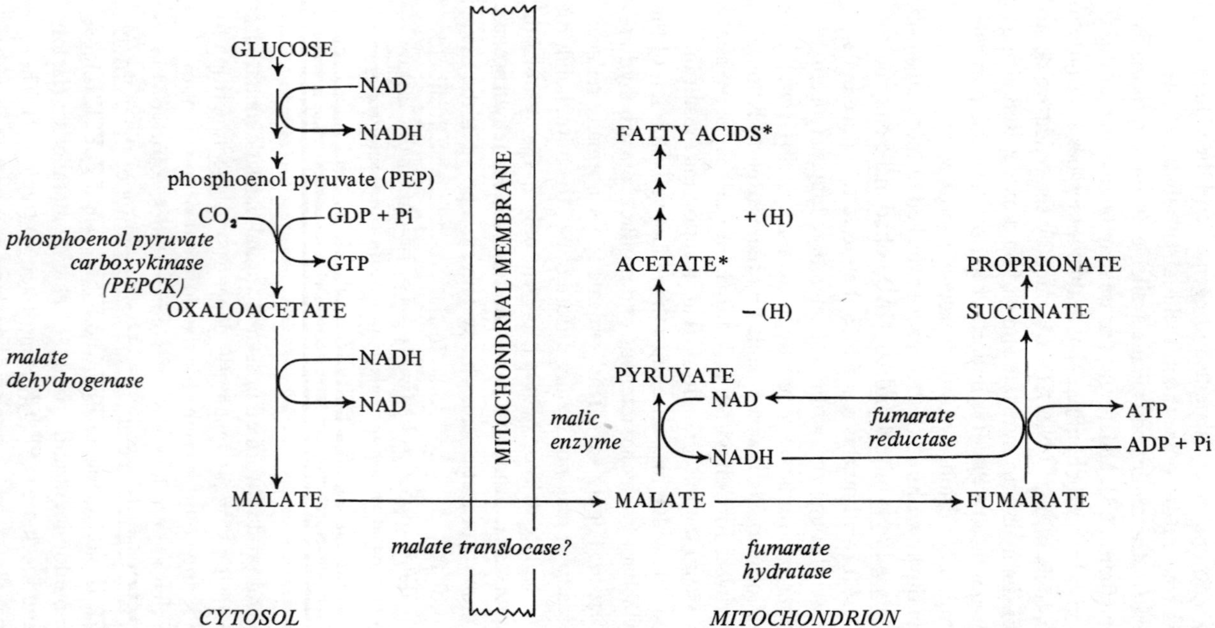

*Pyruvate conversion to acetate and to fatty acids may occur outside the mitochondrion (Bryant[105]).

Figure 4.3 Alternative pathway of carbohydrate metabolism (fumarate reductase pathway) which occurs in *Ascaris* and some other nematodes. (After Saz.[108])

however, that lactate could be produced when PK activity is low if the pyruvate, which is produced from malate within the mitochondrion, were to cross the mitochondrial membrane into the cytosol where lactate dehydrogenase is present.[105]

Table 4.2 Pyruvate kinase/phosphoenolpyruvate carboxykinase activity ratios (PK/PEPCK) in some animal-parasitic nematodes (after Bryant[105])

Nematode	PK activity (nmol/min/ mg protein)	PEPCK activity (nmol/min/ mg protein)	PK/PEPCK	Major end product
Ascaris lumbricoides				
(adult muscle)	7	168	0.04	succinate
Nippostrongylus brasiliensis				
(adult cytosol)	160	86	1.86	lactate + succinate
Dirofilaria immitis				
(adult extract)	453	43	10.6	lactate?
Obeliscoides cuniculi				
(female cytosol)	1182	55	21.5	lactate?
(male cytosol)	1383	78	17.7	
(4th juvenile cytosol)	1098	87	12.7	
(free-living 3rd juvenile cytosol)	2576	39	66.1	
Strongyloides ratti				
(infective juvenile)	69	31	2.2	?
(free-living adult)	56	32	1.8	
(parasitic adult)	7	51	0.14	
(1st juvenile)	40	25	1.6	
Trichinella spiralis				
(juvenile cytosol)	43	140	0.31	succinate
Syphacia muris				
(female cytosol)	6	438	0.01	succinate?
Litomosoides carinii				
(adult cytosol)	137	49	2.8	lactate + succinate + formate

Formation of succinate via the fumarate reductase system produces one more mole of ATP than does the formation of lactate because regeneration of NAD during the formation of lactate is not coupled with the synthesis of ATP as it is during the formation of succinate from fumarate.

This fumarate reductase metabolic pathway is found in several other nematodes, including *Heterakis gallinarum*, *Trichuris vulpis*, *Syphacia muris*, *Dictyocaulus viviparus*, juveniles of *Trichinella spiralis*, and the parasitic female and first-stage juvenile of *Strongyloides ratti*, as well as in several cestodes and trematodes.[105,108] The presence of this anaerobic pathway in the parasitic female and the first-stage juvenile of *Strongyloides ratti* is indicated by the absence of an NAD-dependent, ADP-activated isocitrate dehydrogenase, the very low activity of aconitate hydratase, the low pyruvate kinase activity, the high PEPCK activity (table 4.2) and the absence of key enzymes of β-oxidation. This is in marked contrast to the free-living stages, where all the evidence points to an aerobic metabolism involving the TCA cycle and a cytochrome system of electron transport.[110] The developing egg and first three juvenile stages of *Ascaris* appear to have a normal TCA cycle and are aerobic.[8]

The fumarate reductase pathway may also occur in the free-living nematodes *Ditylenchus triformis* and *Turbatrix aceti*, as fumarate was the predominant organic acid found in these two nematodes.[114]

While the pathway described for *Ascaris* muscle is of frequent occurrence in animal-parasitic nematodes, variants of it do occur. *Dictyocaulus viviparus* apparently has no malic enzyme and, although it possesses most of the enzymes of the TCA cycle, it lacks the oxoglutarate dehydrogenase complex and also fumarate hydratase. It has been suggested that carbon dioxide is fixed via PEPCK into oxaloacetate which then undergoes conversion to malate and fumarate. The fumarate is then converted to succinate by a flavo-protein dehydrogenase and may be excreted or may enter the mitochondrion where it may be oxidised further.[105,118]

4.1.9 The Electron Transport System: Oxidative Phosphorylation

It is generally accepted that the components of the electron transport system are located on the folded, inner membrane of the mitochondrion.

In mammalian tissues the co-enzyme NAD is the first of a series of electron carriers in oxidative phosphorylation (figure 4.4). It is the initial electron acceptor in the oxidation of isocitrate, α-ketoglutarate and malate (figure 4.2).

Three moles of ATP are produced when electrons from NADH pass along the electron transport chain, but only two moles of ATP are produced when the electrons from $FADH_2$ pass along the chain (figure 4.4).

This electron transport system is an aerobic system, but it can operate down to an oxygen tension of about 5 mm Hg within the mitochondrion and possibly even down to 1 mm Hg.[119] It is worth noting here that many nematodes which live in the alimentary tract of animals are closely applied to the mucosa and that oxygen tensions up to 20 mm Hg have been detected in this region of the intestine.[120] It is possible, therefore, that small nematodes that live closely applied to the mucosa may be able to supply oxygen to their mitochondria at tensions greater than 1 mm Hg.

4.1.10 The Electron Transport System in Nematodes

The electron transport system of free-living nematodes has not been extensively studied, but the few examples that have been investigated indicate that the

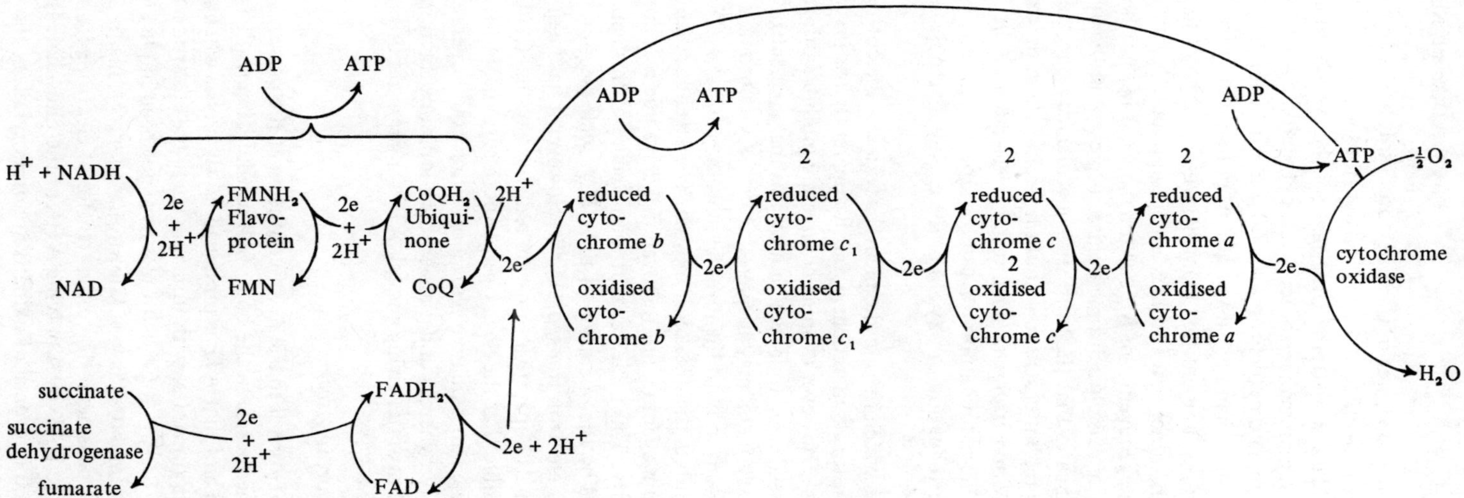

Figure 4.4 Electron transport chain in mammalian tissues.

mammalian type of system operates in these nematodes. It has been shown that mitochondria of *Turbatrix aceti* contain an electron transport system which falls generally into the classical pattern. The peaks for the cytochromes, which are similar to those from mammalian sources, are: cytochrome *b*, 563 μm; cytochromes *c* and c_1, 550 μm and cytochrome *a*, 660 μm. The ratio of cytochrome *c* to cytochrome *a* is greater than in a suspension of mitochondria from mouse liver. Cytochrome a_3 and an unidentified pigment, which is present in much larger quantities than the other cytochromes, are also present.[117] Cytochrome *c* is present in the third-stage juvenile of *Nippostrongylus* but has not been detected in the adult, whereas cytochrome oxidase is present in the adult nematode;[8,103] this indicates that the third-stage juvenile may operate the normal electron transport system, but casts some doubt upon its role in the adult. The cytochromes of adult *Nippostrongylus* need to be investigated further because the dependence of this nematode upon oxygen for motility and survival indicates that there is a direct participation of oxygen in the energy metabolism of this nematode.

Litomosoides carinii, which lives in the pleural cavities of the cotton rat, is dependent upon oxygen for its metabolism, but it apparently has no cytochrome *c* or cytochrome oxidase. However cyanide completely inhibits both respiration and motility of this filarial nematode, which indicates that an electron transport system containing a heavy metal may be present.[108,121] Cytochrome *c* and cytochrome oxidase are present in both the juveniles and the adults of *Trichinella spiralis,* and cyanide inhibits the uptake of oxygen. The juveniles coil up and stop moving in potassium cyanide, but removal of the cyanide is followed by a return to normal motility. This indicates that the juveniles of this nematode can pass electrons produced by the operation of their TCA cycle (section 4.1.7) along a classical type of electron transport chain. However it has been shown that these juveniles can also operate a fumarate reductase system (section 4.1.8).[8,98,105] It is possible that they can operate both systems; in the presence of oxygen the TCA cycle and oxidative phosphorylation could operate, giving enough ATP for motility, whereas under anaerobic conditions the fumarate reductase system could operate, giving enough ATP for survival but not motility.

4.1.11 Electron Transport in *Ascaris*

The situation in adult *Ascaris* is in a state of confusion at the present time. We have seen above that the muscle in the body wall of adult *Ascaris* does not have a functional TCA cycle and that ATP is produced by glycolysis when malate is reduced to succinate via fumarate, with fumarate being the electron acceptor in this system (figure 4.3).[105,108] Cytochromes have not been implicated in this fumarate reductase system, but it is well known that in insect, mammalian and plant mitochondria it is possible to force the transfer of electrons from succinate through cytochrome *b* to NAD by supplying ATP for the phosphate-linked part of the pathway. This is the reversal of oxidative phosphorylation. If the electron flow is from NADH through cytochrome *b* to fumarate, however, then ATP will be synthesised. It is possible that in the fumarate reductase pathway in the muscle of *Ascaris,* the NADH that is produced when malate is decarboxylated to form pyruvate (figure 4.3) will be reoxidised to NAD when two electrons pass first to cytochrome *b,* or possibly to cytochrome *o,* and then to fumarate with the production of succinate. This hypothesis requires testing.

Mitochondria from the muscle of *Ascaris* appear to have cytochromes b_{558}, b_{560}, o, a, a_3, c and c_1. The *b*-type cytochromes are the major component in the respiratory system of these mitochondria and, as suggested above, it is possible that the *b*-type cytochromes or cytochrome *o* may operate in the fumarate reductase system proposed in figure 4.3. However it has been suggested that all of the cytochromes that have been detected in this nematode may be involved in electron transport and, if so, then *Ascaris* must be regarded as a facultative anaerobic organism.[122]

Antimycin A, which inhibits electron transport in mammalian mitochondria between cytochromes *b* and *c,* has no effect on the incorporation of ^{32}Pi into ATP in the mitochondria of *Ascaris,* which suggests that this part of the electron transport chain is not functional in *Ascaris* and supports the fumarate reductase theory.[108] It assumes however that the cytochromes of *Ascaris* are similar to the equivalent mammalian cytochromes and this may not be so.[122] More work is needed on energy production in *Ascaris* as the fumarate reductase system is relatively inefficient when compared to the classical electron transport system, but on current evidence it appears to be the major pathway.

The eggs, first-, second-, and third-stage juveniles of *Ascaris,* unlike the adult nematodes, have an aerobic metabolism and contain both the TCA cycle and the cytochrome *c*–cytochrome oxidase electron transport chain. Cytochrome oxidase appears after the formation of the eight-celled stage during development of the egg; before then a flavin-mediated terminal oxidase is present. Mitochondria from developing eggs catalyse phosphorylation coupled with the oxidation of NADH, the electron acceptor being oxygen, as in mammalian tissues.[8,123]

Trichuris vulpis, although it has a fumarate reductase system, also possesses succinate–cytochrome *c* reductase, cytochrome oxidase and catalase. This nematode consumes oxygen and the activities of the cytochromes are sufficiently high to account for the rate of oxygen consumption.[8] It is possible therefore that this system may play a physiological role in these nematodes, at least under aerobic conditions. This nematode lives closely applied to the mucosa of the host, where a significant amount of oxygen will be present,[120] and this aerobic system may play a more important role in energy metabolism, especially for movement, than the alternative anaerobic pathway, which may maintain a basal metabolic rate but not supply enough energy for motility.

4.1.12 P/O Ratio

Phosphate (Pi) combines with ADP to form ATP during oxidative phosphorylation in mammalian systems. The amount of phosphate involved and the amount of oxygen converted to water can be measured. The ratio of the number of atoms of phosphate incorporated into ATP to the number of atoms of oxygen consumed is the P/O ratio. The P/O ratio for the oxidation of NADH is three, while for succinate it is two. When succinate or pyruvate are the substrate for washed homogenates of *Ascaris* muscle, then the P/O ratio is very low, which indicates that there is no direct coupling between oxidative phosphorylation and the oxidation of succinate or pyruvate in *Ascaris.* Similar results were obtained with *Ancylostoma caninum.* This would be in accord with the fumarate reductase pathway, in which phosphorylation occurs between fumarate and succinate (figure 4.3).[108, 124, 125]

In *Ascaris,* when malate dismutates to half a mole of pyruvate and half a mole of succinate, half a mole of phosphate is esterified into ATP, so that the phosphate/malate ratio is 0.5.[108,124]

4.1.13 The Pentose–Phosphate Pathway

The pentose–phosphate pathway (figure 4.5) is an alternative route to the TCA cycle by which the oxidation of carbohydrates takes place. It usually accounts for only a small percentage of the oxygen taken up by the cell, but at least one of its products, NADPH, is important in cell anabolism. The enzymes concerned are present in the cytosol and not in the mitochondria of the cell. Its primary purpose in most cells is to generate reducing power in the cytoplasm in the form of NADPH, but is also brings about the formation of pentoses, particularly D-ribose, which is used in the synthesis of nucleic acids. The generation of reducing power in the form of NADPH is high in tissues, such as the liver and

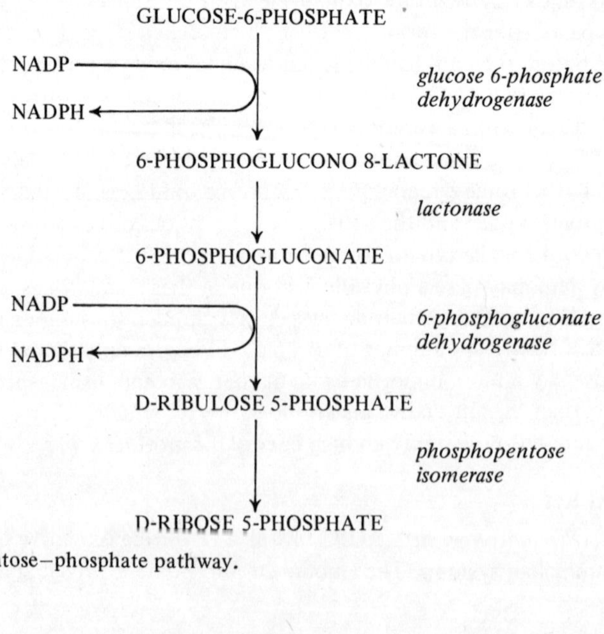

Figure 4.5 Pentose–phosphate pathway.

fatty tissues that carry out the reductive synthesis of fatty acids and steroids, but low in skeletal muscle. The pathway operates in the presence of fluoride ions (the glycolytic pathway is sensitive to fluoride ions) and is relatively insensitive to cyanide ions. The pentose–phosphate pathway is often well developed in lower organisms, but little work has been done on the elucidation of this system in nematodes. Glucose-6-phosphate dehydrogenase and 6-phosphogluconate dehydrogenase are present in several species of animal-parasitic nematodes, in all stages of *Strongyloides ratti* (especially in the free-living generations) and in the plant-parasitic nematode *Ditylenchus.*[106,110,126] It has been suggested that the formation of reduced NADP may be required for lipid synthesis in *Ascaris.* In male *Ascaris* the highest activity of the pentose cycle enzymes is in the muscles of the body wall, while in the female most activity is in the reproductive system.[8,127] This would suggest a correlation with egg production in the female.

4.1.14 Glycogenesis and Gluconeogenesis

Glycogenesis is the synthesis of glycogen from glucose and gluconeogenesis is the synthesis of glucose from non-carbohydrate substrates such as oxaloacetate and lactate.

Much less work has been carried out on glycogenesis and gluconeogenesis in nematodes than on some of the other aspects of carbohydrate metabolism. It is probable that the enzymes involved in both pathways are present in most, if not all, nematodes, as most nematodes contain glycogen. Phosphoenolpyruvate carboxykinase (PEPCK) and fructose-1,6-diphosphatase have been detected in *Panagrellus redivivus,* in *Turbatrix aceti,* in *Ascaris* and in some other nematodes and are probably involved in gluconeogenesis.[8,100,106] However, PEPCK may play a major role in the fixation of carbon dioxide into PEP in *Ascaris.* Pyruvate carboxylase, which converts pyruvate to oxaloacetate during gluconeogenesis, may not be required in *Ascaris* as oxaloacetate can be formed from the succinate that is produced as a result of energy production in the muscles (section 4.1.8). On the other hand it would be required if pyruvate, which had been formed from malate in the fumarate reductase pathway or from the catabolism of amino acids, were to be converted to oxaloacetate.

Of special interest is the fact that the glyoxalate cycle functions in the tissues of some nematodes. Although isocitrate lyase and malate synthase are not normally found in animal tissues they are both found in *Panagrellus redivivus,* in *Turbatrix aceti* and in the eggs of *Ascaris.*[97,116] This will be discussed under lipid metabolism in section 4.2.

4.1.15 Nucleotides and Nucleosides

With the exception of *Ascaris,* little work has been done on the nucleotides and nucleosides of nematodes. The levels of ADP and adenosine monophosphate (AMP) are rather high in the muscles of *Ascaris* and there are measurable quantities of inosine triphosphate (ITP) and guanosine triphosphate (GTP). In this respect they resemble the smooth muscle of vertebrates. The phosphagen content is very low in vertebrate smooth muscle and phosphagens appear to be absent from *Ascaris* muscles. The ATPase of *Ascaris* is of the smooth muscle type and adenylate kinase activity is very high. In some ways, therefore, the biochemistry of the muscles of *Ascaris* is more comparable with vertebrate smooth muscle than with skeletal muscle.[128]

Guanosine and inosine nucleotides are present only in small amounts in the muscle of *Ascaris,* despite the fact that the preferred co-factor for phosphoenol pyruvate carboxykinase is IDP or GDP and not ADP (ADP reacts at less than 25 per cent and GDP at about 75 per cent of the rate of IDP). The muscles of *Ascaris* possess high nucleoside diphosphate kinase activity and this may catalyse the transfer of high energy phosphate from ITP or GTP to ADP, so the flux through IDP or GDP could still remain high in this tissue despite the low IDP/ADP and GDP/ADP ratios. Guanosine nucleotides are present at higher concentrations than inosine nucleotides, which suggests that the flux will be mainly through GDP in this nematode.[128]

NADP is mainly in the reduced form in *Ascaris*, whereas NAD is mainly in the oxidised form. Also there is more NAD present than NADP, which is consistent with the role of NAD in energy metabolism and of NADP in reductive synthetic pathways.[129]

There are two systems which reoxidise NADH in the muscle of *Ascaris*: in the cytoplasm the NADH which is formed during glycolysis is reoxidised by the reduction of oxaloacetate to malate (figure 4.3); in the mitochondria of the muscle the NADH formed when malate is converted into pyruvate by malic enzyme is reoxidised when fumarate is reduced to succinate by fumarate reductase and this also brings about the phosphorylation of ADP (figure 4.3).

4.1.16 Production of Organic Acids

The excretory products of carbohydrate catabolism in adult *Ascaris* and *Ancylostoma* include some unusual volatile organic acids. Succinate is produced by the fumarate reductase system (figure 4.3), but only some of it is excreted. Some of the succinate is decarboxylated to form proprionate, which then forms α-methylbutyrate and α-methylvalerate. The α-methylbutyrate is formed by a condensation of the carbonyl carbon of acetate (which arises from the decarboxylation of pyruvate) with the α-carbon of proprionate (figure 4.3). The α-methylvalerate is formed by a condensation of two proprionate units.[8,108,130]

4.2 Lipid Metabolism

4.2.1 Distribution of Lipids

Lipid is the chief food reserve in free living nematodes, in plant-parasitic nematodes and in the free-living stages of many animal-parasitic nematodes. Glycogen is the chief food reserve in most adult animal-parasitic nematodes but they also contain large amounts of lipid.

Most of the lipid is located in the hypodermis, especially in the lateral cords, in the non-contractile part of the muscle cells, in the intestine and in the ovaries. Lipid droplets are also present in the lumen of the intestine in the free-living, infective-stage juveniles of some animal-parasitic species.

The amount of lipid present in nematodes varies widely (table 4.3), the larger amount of lipid present in most free-living and plant-parasitic nematodes probably being related to their aerobic existence. The differences between the lipid content of nematodes that live under anaerobic or reduced oxygen tensions and those that live under aerobic conditions are shown in table 4.3.

Table 4.3 Total lipids in nematodes from different environments[8,92,109]

Species	Stage	Habitat	Per cent dry weight
Ascaris lumbricoides	Adult	Pig intestine	8.9
Dioctophyme renale	Adult	Dog kidney	1
Eustrongylides ignotus	Juvenile	Fish mesenteries	4.4
Porrocaecum decipiens	Juvenile	Fish muscle	3.5
Nippostrongylus brasiliensis	Adult	Rat intestine	11.9
Trichinella spiralis	Juvenile	Vertebrate muscle	9.1
Strongyloides ratti	Juvenile	Soil/faeces mixture	25
Panagrellus redivivus	Adult	Sour paste	23
Turbatrix aceti	Adult	Vinegar	29
Ditylenchus dipsaci	Adult	Soil (plant parasite)	38
Pratylenchus penetrans	Adult	Soil (plant parasite)	27
Meloidogyne javanica	2nd juvenile	Soil	30
Caenorhabditis sp.	Adult	Soil (microbial feeder)	34–36
Aphelenchus avenae	Adult	Soil (fungal feeder)	32–34

Detailed studies on the nature of the lipids of nematodes have been carried out on relatively few species.[8,96] Most of the phospholipids are probably structural components, while the free fatty acids and the triglycerides probably serve mainly as an energy reserve in aerobic species.

The free-living, non-feeding, infective juveniles of several animal-parasitic nematodes (*Nippostrongylus, Haemonchus, Ancylostoma, Necator*) use their stored lipids as an energy reserve and the infectivity of these juveniles is related to the amount of their stored fat. Juveniles that have depleted their reserves of lipid are not as infective nor as motile as juveniles that contain large amounts of lipid.[48,65] *Tylenchulus semipenetrans, Alphelenchus avenae, Caenorhabditis* sp., *Panagrellus* sp., *Thelastoma bulhõesi* and second-stage juveniles of *Meloidogyne javanica,* use up their stores of lipid if they are starved and, presumably, it is a primary source of energy in these nematodes.[8,92,96,99] Adult *Ascaris*, on the other hand, are unable to utilise their stored lipid to any great extent. Nematodes, including anaerobic species, must have an active lipid metabolism, however, as lipids are deposited in the oocytes and act as a food reserve for the egg. Large amounts of triglyceride are present in the ovaries of *Ascaris* and they are subsequently used by the embryonating egg; the triglycerides disappearing slowly at first and then more rapidly as the egg develops in the presence of oxygen (figure 4.6). The infective juvenile inside the egg uses lipids much more slowly once development is completed. The oxygen consumption actually begins to decrease when lipid is being used most rapidly by the developing egg, which shows that only some of the lipid is being completely oxidised and that the remainder is being converted to carbohydrate (section 4.2.2). As a result the total amount of carbohydrate after twenty-five days at 30 °C is about the same as in the unembryonated egg, but the amount of lipid is greatly reduced (figure 4.6).[97,109]

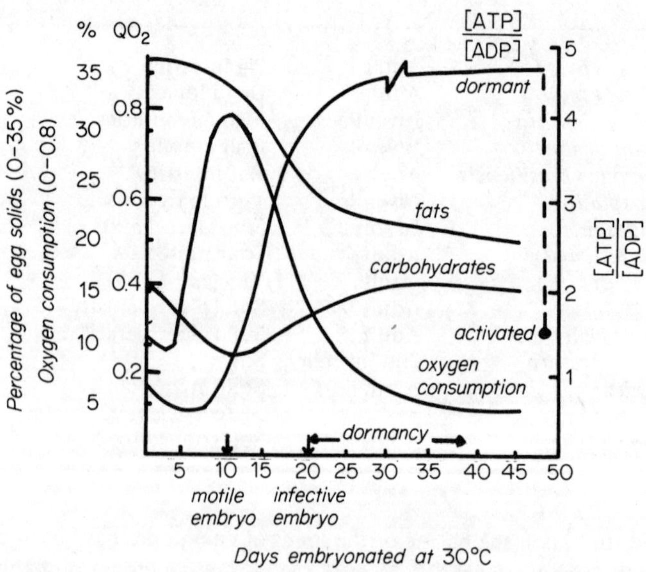

Figure 4.6 Changes in oxygen consumption, lipids, carbohydrates and relative concentrations of ATP and ADP, during development of *Ascaris* eggs. (After Fairbairn;[109] Beis and Barrett.[205])

The oocytes of *Ascaris* also contain large amounts of glycosidic esters in which the high-molecular-weight glycosides (ascarosides) are combined with acetic- and proprionic acid residues. These ascaroside esters appear in the neutral lipid fraction of the ovaries and give rise to the ascarosides that form the ascaroside layer of the egg shell. They are primarily responsible for the extreme resistance of the eggs of *Ascaris*. Fertilisation of the oocyte brings about formation of the egg shell and in this process the ascaroside esters of the oocyte are completely converted to free ascarosides. The ascaroside layer of the egg shell originates from the refringent granules of the oocyte, which presumably contain the ascaroside esters. It has been suggested that this conversion from ascaroside esters in the oocyte and unfertilised egg to free ascarosides in the shell of the fertilised egg is related to the fact that ascaroside esters melt or flow at 40 °C, and an early step in the formation of the ascaroside layer of the shell is the extrusion of the refringent granules and their subsequent coalescence. Once the esters have been converted to free ascarosides the melting point alters to 70–80 °C. It is this which may contribute the stability to this resistant layer of the egg shell of *Ascaris.*[102,103] It is possible that a similar process occurs in the ovaries, oocytes and eggs of several other nematodes, as free and esterified ascarosides have been identified in *Parascaris equorum, Ascaris columnaris, Ascaridia galli, Toxocara cati* (all ascarids) and in the oxyurid *Passarulus.* These, or closely related compounds, have also tentatively been identified in two other oxyurids and in a free-living rhabditid. They have not, however, been detected in the free-living nematodes *Turbatrix aceti* and *Panagrellus redivivus.*[101]

4.2.2 Lipid Metabolism

It appears that the ability to synthesise sterols is absent in nematodes as the free-living nematodes *Panagrellus redivivus, Caenorhabditis briggsae* and *Turbatrix aceti,* and the animal-parasitic nematode *Ascaris,* are unable to synthesise sterols from mevalonate or acetate.[92,131,132,133] There must however be a rapid turnover of lipid, especially in those nematodes such as *Ascaris,* in which there is a massive production of eggs.

Ascarylose, which is important in the production of the egg shell, is synthesised in the ovaries of *Ascaris* from either glucose or glucose-l-phosphate by a sequence of reactions that probably involves a nucleotide-diphosphate glucose as an intermediate, and NAD and NADPH as co-enzymes. The end product of synthesis is probably ascarylose-dinucleotide-phosphate, which then condenses with aglycones to form ascarosides. The aglycones are formed from a reaction between two moles of palmitic acid or another higher fatty acid.[8,103] Large amounts of triglyceride are present in the oocytes of *Ascaris* and of other nematodes, and these are incorporated into the egg to be used in embryogenesis when the egg is passed out of the host. The origin of these triglycerides is unknown, but it has been suggested that they may arise, via fatty acids, as products of anaerobic carbohydrate metabolism. It is worth noting here that adult *Nippostrongylus*, when at the peak of egg production in the host, have about 11 per cent of their dry weight as lipid and that the major, neutral lipid fraction is the free fatty acid fraction. When the immune response of the host drives these nematodes away from their aerobic environment between the villi of the intestine into the lumen of the intestine (an anaerobic environment, or nearly so) there is a rise in the total lipids and a change from free fatty acids to triglyceride as the major component of the neutral lipids. This indicates that under aerobic conditions the nematode has an active lipid metabolism, but under anaerobic conditions this slows down or ceases and free fatty acids that are produced are directed into triglycerides. On the other hand, it could be that when egg production ceases in the nematodes affected by the immune response of the host, the triglycerides, which would normally pass out with the eggs, are stored in other tissues as the nematodes are not able to catabolise them.[85,134]

Lipids are used extensively by free-living nematodes and by the free-living juveniles of animal-parasitic nematodes, which presumably have a fairly normal lipid metabolism.[48,65,92,99]

The breakdown of endogenous fats to release energy requires oxygen in nematodes, as in other animals. Juveniles of *Trichinella spiralis,* when kept in non-nutrient media under anaerobic conditions, do not use any of their stored lipid, but when oxygen is admitted there is a significant decrease in the amount of stored lipid and this is independent of the carbohydrate consumed.[8] Similarly, *Aphelenchus avenae* uses its lipid when starved under aerobic conditions but not under anaerobic conditions.[99]

The unembryonated eggs of *Ascaris* contain large amounts of triglyceride. The amount decreases slowly during the first five days of embryonation, then rapidly until the twentieth day, during which time the embryo has developed into a second-stage, infective juvenile. From then on the triglycerides disappear slowly. At the period of embryonation when there is extensive depletion of trigylcerides, there is extensive resynthesis of glycogen and trehalose and a decrease in oxygen consumption (figure 4.6). This suggests that there is conversion of fatty acids to carbohydrate. Apparently, while some of the triglyceride is oxidised completely to produce energy, some is oxidised as far as acetyl CoA and then this is used to synthesise carbohydrate by β-oxidation of fatty acids via the glyoxylate cycle. In this cycle (figure 4.7) acetyl CoA, which arises from the β-oxidation of fatty acids, reacts with glyoxylate to form malate, which is glycogenic. The two enzymes involved in the glyoxylate cycle are isocitrate lyase and malate synthetase. Both of these enzymes appear at an early stage in embryonation of the eggs of *Ascaris* and are most active when the rate of fatty acid oxidation and carbohydrate synthesis is at its peak. There is, therefore, no doubt that a mechanism exists and operates in the developing egg of *Ascaris* for the conversion of lipids to carbohydrates.[97] The enzymes of the glyoxylate cycle have also been shown to occur in the free-living nematodes *Caenorhabditis, Panagrellus, Turbatrix* and *Rhabditis* and, presumably this cycle is functional in these nematodes.[92,116]

4.3 Protein Metabolism

4.3.1 Distribution of Proteins

Insoluble proteins are the chief structural components of the cuticle, muscles and other tissues of nematodes. Soluble proteins, such as glyco- and lipo-

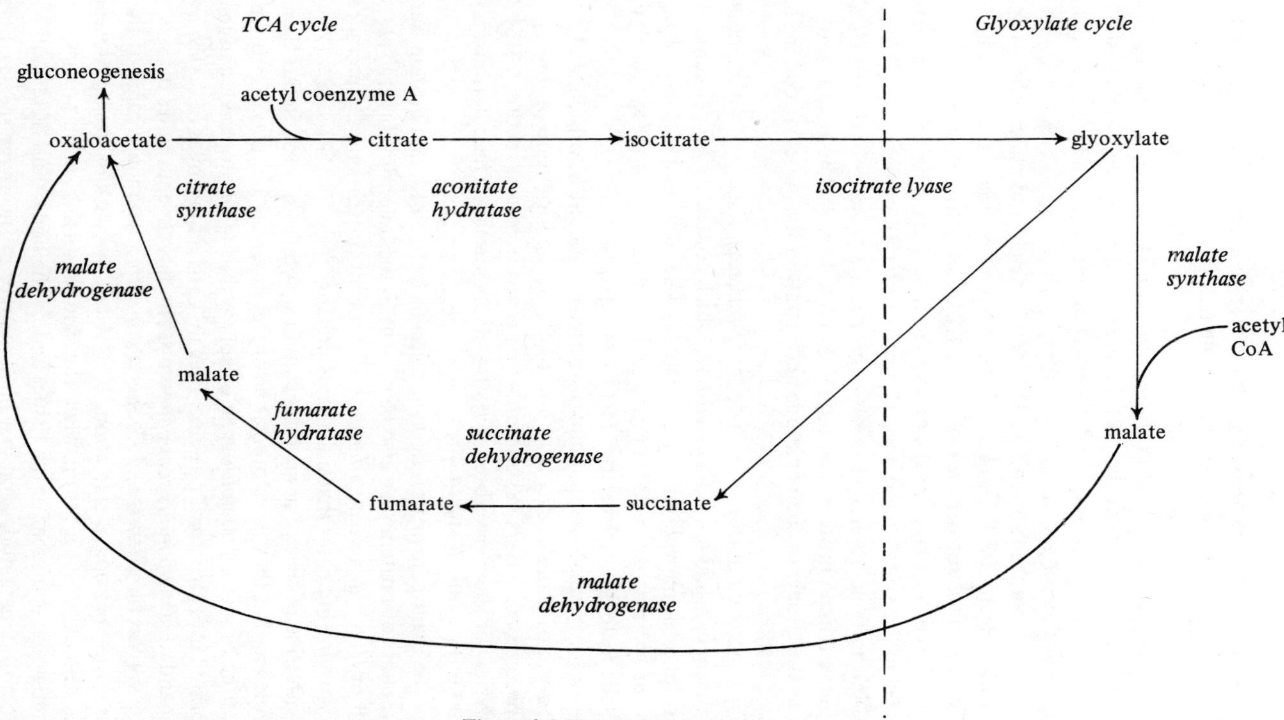

Figure 4.7 The glyoxylate cycle.

proteins, occur in the various tissues and in the body fluid. The total protein content is known for relatively few nematodes, but it varies between about 50 and 80 per cent of the dry weight of the tissues. The apparent differences in the percentage protein content of various nematodes may be due partly to variation in the amounts of carbohydrate and lipid in the tissues. Soluble proteins have been used for taxonomic purposes, as distinct differences in pattern occur between species when the proteins are separated by electrophoresis or isoelectric focussing. The soluble proteins of various stages may also vary during development of a species.

Collagen is present in the cuticle, muscles, egg shell and a few other structures of nematodes (section 2.1.2 and 6.1.6). An unusual protein called cuticlin is also present in the cuticle of *Ascaris* (section 2.1.2).[30] Keratin-like proteins may be present in the cortical layer of the cuticle, in the cuticular lining of the pharynx and cloaca, in the spicules and possibly in the stylet, of nematodes.[4,5] After the death of gravid female *Heterodera rostochiensis* the proteins in the cuticle alter into quinone-tanned protein. The tanning is brought about by polyphenol oxidase in the cuticle of the nematode and results in the cuticle becoming a tough cyst wall.[31]

The spermatozoa of several nematodes (*Parascaris, Ascaris*) contain granules of an unusual compound called ascaridine, which contains 17.5 per cent nitrogen but no sulphur or phosphorus. These granules fuse to form a central, refringent body in the spermatozoa and they play an important role in the synthesis of RNA in the male pronucleus after fertilisation (section 6.1.5).

The ovary of nematodes contains a significant amount of protein and some is in the form of granules that later contribute to the formation of the egg shell. The egg shell differs in structure from genus to genus (section 6.1.6), but relatively few of them have been investigated chemically. The outermost layer is secreted by the walls of the uterus in some species and appears to be a mucoprotein. The main part of the egg shell contains both chitin and protein but small amounts of protein may also be present in the innermost layer. The egg shell of *Heterodera rostochiensis* is 59 per cent protein and the presence of hydroxyproline indicates that collagen is present.[135]

The intestine of several nematodes contains globules of protein but there is little evidence that they are stored as an energy reserve. Granules of protein in the intestine of juvenile *Meloidogyne* disappear during development and may be incorporated into structural proteins. Freshly hatched juveniles of *Meloidogyne javanica* and of *Tylenchulus semipenetrans,* and infective juveniles of *Nippostrongylus brasiliensis* and *Cooperia punctata,* metabolise some of their proteins during periods of starvation.[8,92,96] Adult mermithids use up protein granules stored in their intestine when they are producing sperm or eggs.

Haemoglobin is present in the tissues and/or the body fluid of several animal-parasitic nematodes and in a small number of free-living species. Some nematodes contain more than one haemoglobin, for example *Ascaris* contains one haemoglobin (possibly two) in the body wall and one in the pseudocoelomic fluid, while *Obeliscoides cuniculi* contains three haemoglobins in the body wall, two in the pseudocoelomic fluid and one in the intestinal wall.[136] Haemoglobin from the body wall of *Ascaris* has a molar mass of 37 000, whereas the value of that of the haemoglobin from the pseudocoelomic fluid is 328 000.[137] The haemoglobins of nematodes have a high affinity for oxygen and in some nematodes the haemoglobins function in an oxygen-carrying or oxygen-storing capacity (section 4.4.5). This is, apparently, not so for all nematode haemoglobins and various functions for these pigments have been suggested. They may play a role in the elimination of peroxide or they may be a source of haem for other haemoproteins that are incorporated into the egg.[83]

4.3.2 Protein Metabolism

There must be extensive metabolism of amino acids and proteins during growth, moulting and egg production in nematodes. Much of the nitrogen metabolism in female nematodes is concerned with the formation of the eggs, which is not surprising when it is realised that a single female *Ascaris* can lay 200 000 eggs a day and that many other animal-parasitic species are equally prolific. Most of the proteins deposited in the oocyte become structural components for the egg shell of the juvenile nematode and are not used as a source of energy.

Many organisms are able to convert ammonia to α-amino acids but this activity is commonest in plants and micro-organisms since most animals rely on dietary amino acids for protein synthesis. Ammonium salts, such as ammonium chloride, are converted to glutamic acid by glutamic acid dehydrogenase. This enzyme is situated mainly in the mitochondria where α-ketoglutarate, which is an acceptor of the ammonia, is also present. An imino acid is formed from α-ketoglutarate and ammonia and this is then reduced to glutamic acid by electrons from NADH or NADPH. This glutamic acid can then be incorporated directly into proteins, or the amino group of glutamic acid can then be transferred to α-keto acids, such as α-ketoglutarate, pyruvate or oxaloacetate, converting them into α-amino acids. The enzymes that bring about this transformation are called amino transferases. *Ancylostoma canimum* contains glutamic acid dehydrogenase and glutamic-acid–pyruvate aminotransferase and probably carries out synthesis of amino groups in this way.[138] If the reaction is reversed, then ammonia is removed from the α-amino acids and eventually α-ketoglutarate and ammonia are produced. This ammonia must be excreted or transformed into other nitrogenous compounds, as it is toxic to cells.[8,139]

Whole female *Ascaris* contain a glutamic acid dehydrogenase which has a very high affinity for ammonium ions (2000 times greater than beef liver) and glutamic acid is produced as described above.[140] The ovaries of *Ascaris* can synthesise alanine and aspartic acid by reductive amination if pyruvate and ammonia are present.

$$NH_3 + \text{pyruvate} + NADH_2 \rightarrow \text{alanine} + NAD$$

If *Ascaris* is maintained *in vitro* the high levels of ammonia and amino acids in the ovaries are maintained only if there is ammonia in the incubating medium. Under anaerobic conditions the formation of alanine depends upon the presence of NADH and ATP. The continuity of the process depends upon the removal of the alanine from the system and this is brought about by transamination.[141]

The ovaries contain an alanine α-ketoaminotransferase, but the corresponding glutamic-acid-α-ketoaminotransferase is not present. Alanine, rather than glutamic acid, is apparently the central point of amino acid metabolism in the ovaries of *Ascaris.* Alanine–glutamic-acid and aspartic-acid–glutamic-acid aminotransferases are present in the ovaries and the muscles of *Ascaris.* Transaminations involving alanine, valine and methionine are most active in the intestinal cells and least active in the muscles of *Ascaris.* Transaminations which produce glutamic acid or alanine with α-ketoglutarate or pyruvate occur in *Dictyocaulus filaria.*[8] Alanine can be formed by transaminations involving aspartic acid or glutamic acid and pyruvate in the plant-parasitic nematode *Aphelenchoides ritzemabosi.*[142] Pyridoxal phosphate, which is a derivative of vitamin B_6, is an essential co-factor in transaminations and several nematodes have been shown to have a requirement for it in their diet.[92]

Amino acid decarboxylases catalyse reactions in which carbon dioxide is removed from an amino acid to leave a primary amine. These reactions also require pyridoxal phosphate as a co-factor. Glutamic acid decarboxylase is present in *Ascaridia galli* and several decarboxylases are present in the tissues of *Ascaris,* since carbon dioxide is released when amino acids are added to homogenates of this nematode.[8] Amino acid carboxylases are involved in the formation of a number of important compounds. For example, serine is decarboxylated to form ethanolamine, which is a precursor of choline; decarboxylation of cysteine and aspartic acid form β-thioethylamine and β-alanine, which contribute to the structure of coenzyme A, an important component in the formation of acetyl co-enzyme A for the TCA cycle. This aspect of amino acid metabolism has been little studied in nematodes.

Amino acids taken in by nematodes during feeding are readily incorporated into proteins. Juveniles of *Mermis nigrescens* rapidly synthesise proteins from amino acids which are present in the haemolymph of the insect host but these amino acids are incorporated at different rates. Glucose is also metabolised and incorporated into proteins.[87] Amino acids are an important dietary source of protein nitrogen for juveniles of *Trichinella spiralis* and microfilariae of *Dirofilaria immitis.* Amino acids have to be supplied in the culture media when nematodes are grown *in vitro,* but some nematodes ingest particulate protein and presumably break this down into amino acids in the intestine (section 3.3.6).[8,92,93]

Most animal-parasitic and many plant-parasitic nematodes live in environments that provide an abundant supply of amino acids but species that live in soil, fresh water or seawater may be more restricted in their intake of amino acids. It is generally accepted that animals do not have the ability to synthesise several amino acids and these 'essential' amino acids must be obtained from their food. *Caenorhabditis briggsae,* however, can synthesise several of these 'essential' amino acids. When maintained in water containing acetate-2-^{14}C, glycine-1-^{14}C or glycine-2-^{14}C the nematodes synthesised threonine, tyrosine, valine, leucine, isoleucine, histidine and lysine as well as several 'non-essential' amino acids. *Meloidogyne* and *Ditylenchus reformis* can produce several amino acids when incubated in phosphate buffer containing acetate-2-^{14}C, but only *Meloidogyne* contained a labelled 'essential' amino acid.[92]

Little is known about the incorporation of amino acids into proteins in nematodes. The collagens that occur in the cuticle and muscles of *Ascaris* differ in their molar mass and in their amino acid content. Cuticular collagen has a molar mass of 900 000 (the molar mass of vertebrate collagen is 300 000) and contains amino acids in slightly different amounts to vertebrate collagens, elastin, fibrin, resilin or keratin (table 2.1) (see section 2.1.2 for further details of cuticular collagens). Collagen from the muscles of *Ascaris* contains more hydroxyproline, hydroxylysine and carbohydrate than cuticular collagen. It is of interest to note that protocollagen hydroxylase from the somatic muscles of *Ascaris* functions most efficiently at oxygen concentrations between 1 and 4 per cent and is inhibited by 10 per cent oxygen. Activity of this enzyme from the fluid between the cuticle and the muscle is increased at oxygen concentrations higher than 10 per cent and it is suspected that there are different isoenzymes in the muscle and the hypodermis. Protocollagen hydroxylases from the muscles and from the egg of *Ascaris* are different isoenzymes and the egg isoenzyme is not inhibited by concentrations of oxygen higher than 5 per cent. It would appear that the isoenzymes of the adult body wall and of the egg are adapted to the oxygen requirements of these different stages and also to the oxygen tensions in the environment.[8,33,143]

Ascaris is able to digest ingested mammalian haemoglobin to release haematin and amino acids, some of which are then incorporated into either body wall or pseudocoelomic fluid haemoglobin. Body wall and pseudocoelomic fluid haemoglobins consist of the same amino acids but the amino acid residues differ. Pseudocoelomic fluid haemoglobin consists of eight sub-units, each of which has a molar mass of 40 600 whereas the body wall haemoglobin has a molar mass of 37 000 and resembles a single unit of pseudocoelomic fluid haemoglobin.[8,83] Excretion of the end-products of nitrogen metabolism is dealt with in section 5.3.4.

4.4 Respiratory Physiology

4.4.1 Diffusion of Oxygen

Nematodes lack respiratory and circulatory organs and their aerobic metabolism must rely on the diffusion of oxygen into their tissues from their surroundings. Most nematodes have a cylindrical body shape and the diffusion pathway is from the body surface through the body wall and pseudocoelomic fluid to the gut. As the oxygen diffuses into the animal it is utilised by the tissues and this forms a diffusion gradient with a higher partial pressure of oxygen at the surface of the animal than in its central tissues. The major factors governing the adequacy of this diffusion process at a given level of aerobic metabolism are the oxygen tension in the niche and the body radius of the animal. Many species will ingest oxygen with their food and in some cases (*Ancylostoma, Ascaris*) this may represent a significant source of oxygen diffusing outwards through the tissues from the alimentary tract of the nematode. Partial pressures of less than a few millimetres of mercury are adequate for oxidative phosphorylation within the mitochondria of several organisms.[119] If this also applies to the mitochondria of nematodes then a partial pressure of 15 mm Hg at the surface of a species with a maximum body radius of less than 0.1 mm should provide an adequate supply of oxygen to all their tissues. This generalisation cannot be rigidly applied because a number of other factors are also involved. Many animals do not live in a homogeneous medium with a constant partial pressure of oxygen at the surface of the nematode. The different tissues of a nematode probably have different oxygen requirements and not all are likely to be equally permeable to oxygen. The diffusion gradient in many species may be altered by locomotory movements of the nematode as they will cause mixing and movement of both the pseudocoelomic fluid and the fluid surrounding the animal.[48,144] *Ascaris* is of particular interest because it has an exceptionally large body radius and it inhabits a low-oxygen regime. It may not be as dependent on aerobic metabolism as most other nematodes (section 4.1.8), however its haemoglobin may increase the rate of diffusion of oxygen into the animal (section 4.4.5). This facilitated diffusion, and the reduction of the diffusion gradient in the pseudocoelomic fluid by mixing, could enable a significant aerobic metabolism to be maintained at quite low partial pressures of oxygen in spite of its large body size.

4.4.2 Factors which Influence Oxygen Demand

The oxygen requirements of an aerobic nematode are related to its level of metabolism and are influenced by a number of endogeneous factors such as body size, and by environmental influences, on oxygen uptake. It is widely accepted that the rate of respiration of most animals decreases with increasing body size. Where a number of species that have a wide range in body size are considered oxygen consumption is proportional to 0.75-power of the body weight (figure 4.8). At present there is no reason to assume a different value for nematodes. The factors governing this value in animals are poorly understood. It may represent an evolutionary compromise between the advantages of maintaining a high level of metabolism with increasing size and the limitations imposed by the reduced effectiveness of surface-dependent processes in larger animals. The important consequence of this relationship is that comparisons of the rate of metabolism of two nematodes should make allowance for any difference in their body size. A large nematode such as *Ascaris* will probably maintain a much lower rate of metabolism per unit body weight than a much smaller species such as *Panagrellus.*[144]

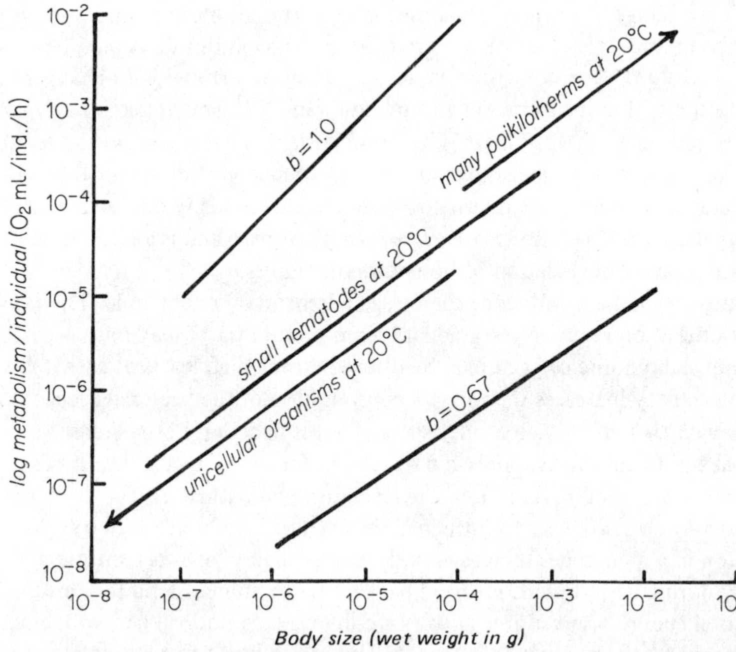

Figure 4.8 The relationship between the metabolism (M) of invertebrate animals expressed as O_2 consumption and body size measured as wet weight (W). The linear relationship after logarithmic transformation expresses a simple exponential formula.

$$M = kW^b$$

where b and k are constants.
b is the slope of the logarithmic line and k determines its elevation at a given body weight. $b = 0.75$ in most poikilotherms, homiotherms and nematodes ($b = 1.0$, indicating a weight-independent metabolism and $b = 0.67$, generally taken to indicate that metabolism is proportional to surface area, are given for comparison). Available data suggest that k for small nematodes has a value between that for unicellular organisms, and the higher value typical of many larger poikilotherms. (Data from Atkinson.[144])

The oxygen consumption per unit weight may also fall with increasing body size during growth of a single species. In this case a variety of exogeneous and endogeneous factors may act in a species-specific way and cause intraspecific values for the weight exponent, b (figure 4.8) that differ significantly from $b = 0.75$. In *Panagrolaimus rigidus* a constant value of $b = 0.64$ is maintained throughout development,[145] but this value may not occur in all nematodes. A large modification in value for b is associated with a change from a non-infective to an infective juvenile in *Nematospiroides dubius.*[146] This intraspecific weight exponent may also be influenced by environmental changes. At very low oxygen tensions small males of *Enoplus brevis* consume more oxygen in total than large individuals. This could be explained if the larger males lapse more readily into quiescence when the amount of oxygen in the environment is low.[111]

Changes in temperature alter the metabolic rate of animals. In many nematodes the rate of change in oxygen consumption with alterations in temperature is probably a more or less constant increment, but exceptions do occur. The uptake of oxygen by *Ditylenchus dipsaci* varies in a complex manner with changes in temperature within its normal environmental range.[147] The uptake of oxygen by infective juveniles of *Nippostrongylus brasiliensis* is enhanced when the temperature approaches that of the skin of its host to a much greater extent than normally occurs in animals. This thermal response occurs in many infective juveniles of skin-penetrating nematodes (*Necator, Ancylostoma*) and is a physiological adaptation favouring location of homeothermic hosts.

Changes in activity influence the oxygen demand of a nematode. The rate of energy utilisation required for gradual locomotion in water may only be slight, but appreciably more oxygen may be utilised during high levels of activity. Feeding activity increases the oxygen consumption of the bacterial feeder *Pelodera chitwoodi.*[148] Nematodes show changes in activity in response to external stimuli and the suppression of locomotory activity at other times may also involve a general reduction in the level of metabolism.

Starvation is also likely to influence the level of metabolism and oxygen consumption in a nematode. Increased body size probably favours continued activity during periods without food because larger animals tend to consume their total energy reserves at a slower rate than smaller individuals, which have a higher metabolic rate per unit weight. This seems likely to be correct for nematodes, and larger free-living forms may show a greater tolerance of starvation than smaller species.

Many environmental factors may influence the oxygen consumption of a nematode in either a direct or indirect manner, for instance changes in oxygen consumption with osmotic pressure are not solely due to the energy required for ionic regulation. This stress may also influence the muscle tone of the animal and alter the level of activity.

4.4.3 Availability of Oxygen

The availability of oxygen is the most important environmental modifier of oxygen consumption in some nematodes. The atmosphere is the ultimate source of oxygen and nematodes are dependent on the replacement of dissolved oxygen in their fluid environments by diffusion from air or from the blood of their hosts. In some habitats this process fails to ensure that the individual experiences a constant level of oxygen supply comparable with that of aerated water.

The intestinal lumen of homeotherms is the most fully examined low-oxygen regime inhabited by nematodes. There is a marked oxygen gradient from the bloodstream, through the host tissue and mucosa, into the lumen of the alimentary tract. Virtually no oxygen is available to a nematode in the centre of the lumen as many of the micro-organisms present use the gas and reduce its concentration to a very low value. The amount of oxygen that is present close to the mucosa is appreciable, being about 25 mm Hg in the small intestine of the duck.[120] Many intestinal nematodes maintain a position against the wall of the alimentary tract. *Nippostrongylus brasiliensis* remains close to the mucosa, frequently lying between the villi, and *Trichostrongylus colubriformis* burrows beneath the mucus and superficial layers of the mucosa.[149] Nematodes with a small radius are able to maintain an aerobic metabolism when close to the intestinal wall. The ascarids are much larger than *Nippostrongylus* and *Trichostrongylus* and are less able to maintain such intimate contact with the mucosa. It is widely believed that there is not sufficient oxygen available to *Ascaris* for it to support an appreciable aerobic metabolism. It has also been argued that the animal relies on oxygen derived from intermittent contact with the mucosa to raise its level of aerobic metabolism within the gut lumen.[150] It would appear that *Ascaris* does not receive a constant supply of oxygen to all of its tissues but oxygen may be available to much of the animal when it is braced against the intestine of its host or ingests food derived from close to the surface of the mucosa.

Marine- and fresh-water muds also show a gradient of oxygen concentration from that of the water at the surface to anoxic conditions at a variable depth within the mud. Some nematodes have been recovered from apparently anaerobic muds. For example nematodes occur at a considerable depth in an African swamp, but little is known of their physiological adaptations to this environment.[144] Few soils are normally anoxic but many show seasonal changes in the oxygen content of the soil atmosphere. Soil moisture content is particularly important because the rate of oxygen diffusion through air is several thousand times greater than its rate through water. Therefore in a water-logged soil the rate of oxygen diffusion from the air is unlikely to equal the oxygen demand of the soil organisms. Micro-organisms have a smaller body radius than nematodes and are liable to reduce the amount of oxygen in the water to a level below that required by nematodes for an optimum level of respiration. Nematodes in soil are therefore liable to fluctuations in oxygen availability with periods of anoxia possible in those soils susceptible to waterlogging.

4.4.4 Respiration in Low-oxygen Regimes

Lack of oxygen suppresses the over-all metabolism of many nematodes and influences a number of different activities. In many species the rate of development is suppressed by low oxygen concentrations (*Aphelenchoides composticola, Ditylenchus myceliophagus, Ascaris*).[144] The effect of low-oxygen regimes on the rate of hatch of *Meloidogyne* is probably typical of many nematodes[151] (see figure 6.9). *Ascaris, Ascaridia* and *Toxocara* are interesting exceptions because they hatch in virtually anoxic conditions. This would appear to be an adaptation to parasitism.

Most, if not all, nematodes require some oxygen during locomotion. Anoxic conditions impair movement in *Caenorhabditis, Nippostrongylus* and *Meloidogyne* but not all species are readily killed by these conditions. Inactivity of some species may result from an inadequate anaerobic energy metabolism. However this is not necessarily the case, for the nerves of many animals are susceptible to lack of oxygen. Some species may remain active in an inadequate oxygen regime by using anaerobic pathways, just so long as sufficient oxygen is available for the aerobic requirements of a few tissues, such as their nerves.

The over-all effect of lack of oxygen on *Caenorhabditis* and *Aphelenchus avenae* can be seen from its effect on the rate of growth of the population. Fluctuations in the amount of oxygen to levels less than 35 mm Hg suppress the increase in numbers of these species. The interruption in catabolism caused by transient lack of oxygen apparently involves some metabolic inertia, which results in a markedly disproportionate reduction in growth of the population.[152] The metabolism of non-feeding, infective juveniles of animal-parasitic nematodes may also be reduced by low-oxygen regimes and longevity may be enhanced in conditions that are unsuitable for host invasion.

It is clearly established that fluctuations in the supply of oxygen influence the metabolism of many nematodes. It is now necessary to consider whether nematodes possess adaptations that reduce the limitations imposed by low partial pressures of oxygen.

The oxygen consumption of a nematode may fall with decreasing ambient oxygen concentration (*Ascaris*) or may remain more or less constant over at least a part of a range of partial pressures of oxygen (*Nippostrongylus, Haemonchus, Enoplus communis*). Interpretation of the precise relationship between oxygen consumption and oxygen availability is complex. A reduced rate of diffusion into a small nematode only limits consumption of oxygen at very low oxygen tensions. It has been suggested that *Enoplus* maintains a lower level of metabolism at low oxygen tensions than that imposed by the rate of diffusion. This would be of homeostatic value because the level of metabolism would not be so readily influenced by fluctuations in oxygen availability. A somewhat lower, but constant, aerobic metabolism may be preferable to a slightly higher level of oxygen consumption that is subject to frequent interruptions in the supply of oxygen. It may be incorrect to consider the oxygen consumption of a nematode in aerated water as indicative of the optimum level of metabolism for a species that normally inhabits an environment where the oxygen tension is low. A more useful approach in the future may be to find out if the aerobic metabolism of a species at an oxygen tension that is typical of its environment is adequate to support the range of activity normally occurring in its niche.[144] *Mesodiplogaster lheritieri* shows a different and interesting adaptation to a low-oxygen regime. This nematode swallows air at the surface of fluids with a low oxygen content and this source of the gas may contribute significantly to the uptake of oxygen by the animal.[153]

Aerobic metabolism will be limited even in small nematodes when the supply of oxygen is so low that it approaches anoxia. Many nematodes become inactive and lapse into a quiescent state in such conditions (*Nippostrongylus, Enoplus communis, Caenorhabditis, Aphelenchus*). Under these circumstances anerobic pathways in *Caenorhabditis* maintain a low level of metabolism but the animal becomes quiescent, which suggests that anaerobiosis cannot entirely compensate for lack of oxygen in many nematodes that become inactive. Many species die if they are subjected to several hours of anoxia (*Enoplus communis, Caenorhabditis, Nippostrongylus*) but *Aphlenchus* descends into a cryptobiotic state. Under these circumstances the animal does not have a detectable metabolic rate and this suspended condition is associated with a longer period of survival.[92,99]

It is possible that anaerobic pathways may be exploited more fully by certain animal-parasitic nematodes when there is a low oxygen tension in their environment. An extreme example may be *Ascaris,* which some workers believe to be virtually independent of oxygen for the physiological production of energy *in vivo* (section 4.1.8).

4.4.5 Haemoglobin

Haemoglobin occurs in many invertebrates and is, undoubtedly, an important factor in the respiratory homeostasis of those species that possess it. Many animal-parasitic nematodes contain haemoglobin. The free-living adults of *Mermis* also contain a little haemoglobin but *Enoplus brevis* is the only free-living nematode known to contain appreciable quantities of the pigment.[144,157] Both this species, from marine mud, and the animal-parasitic forms that contain haemoglobin, are larger than many nematodes yet they inhabit environments with low oxygen tensions. The presence of haemoglobin in these animals suggests that it is associated with ensuring an oxygen supply in nematodes that have both a relatively long diffusion pathway to their internal tissues and a low oxygen concentration in their habitat. The pigment is rarely distributed equally throughout all the tissues of a nematode and may primarily serve specific tissues rather than enhance the oxygen supply to all parts of the animal.

The haemoglobins of animal-parasitic nematodes and their hosts have different properties, which indicates the presence of a haemoglobin metabolism in these parasites. In *Ascaris,* mammalian haemoglobin can be broken down and absorbed, resulting in an increase in the amount of *Ascaris* haemoglobin present in the pseudocoelomic fluid.[150] Certain animal-parasitic nematodes have a number of distinct haemoglobins with different properties. The haemoglobin in the pseudocoelomic fluid of *Ascaris* has the highest oxygen affinity of any known animal haemoglobin. A similar pigment occurs in *Strongylus* and both animals appear to be unable to deoxygenate the pigments in their pseudocoelomic fluid.

These particular haemoglobins may not have a respiratory function; that in *Ascaris* may represent a pool of haematin intended for transfer into the gametes of the parasite for the formation of haemoproteins.[154]

There is little doubt that many of the tissue haemoglobins of nematodes are respiratory pigments. Those isolated from parasites of the alimentary tract have a number of common characteristics. They have a molar mass of 37 000 and have one haem group per protein molecule as found in the myoglobin rather than the haemoglobin of mammals.[83] *In vitro,* the haemoglobins of nematodes have high affinities for oxygen and are half saturated with oxygen at partial pressures between 0.05 mm Hg and 9.0 mm Hg (figure 4.9). Those species investigated also have a much lower rate of dissociation from oxygen than either

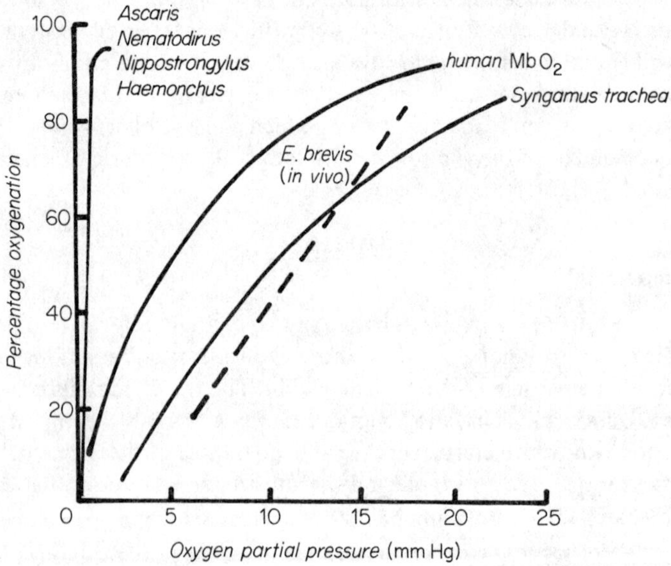

Figure 4.9 Approximate comparison of the oxygen affinity of some animal-parasitic nematodes *in vitro* and the marine nematode *Enoplus brevis in vivo* with human myoglobin (MbO_2). (Partly after Fairbairn[109] and Atkinson.[157])

the haemoglobins or myoglobin of mammals. There is some evidence that nematode haemoglobins are adapted to the specific requirements of certain tissues for oxygen. A high haemoglobin content is frequently found near neural tissue and this may be associated with the limited ability of nerve cells to function under anaerobic conditions. Haemoglobin is also present in the fluid layer of the cuticle of adult *Nippostrongylus* and probably aids the inward diffusion of oxygen from the oxygen-rich mucosa to the body tissues.[5]

The high oxygen affinity of the haemoglobin from nematodes that inhabit the alimentary system may be related to a respiratory function *in vivo* under low oxygen tensions. In contrast, the pigment of *Syngamus trachea* deoxygenates *in vitro* at a higher oxygen partial pressure. This may be correlated with a respiratory function in the higher partial pressures of oxygen in the trachea.[155] Its presence in this nematode may be associated with the difficulties of supplying oxygen to a large species with a correspondingly long diffusion pathway to the internal tissues.

It is believed that one of the main functions of invertebrate tissue haemoglobins is to facilitate the diffusion of oxygen through tissues. Under optimal conditions facilitated diffusion can increase the rate of diffusion of oxygen several fold. It makes its greatest relative contribution to the oxygen supply of a tissue when the oxygen gradient causes the haemoglobin to deoxygenate near the site of utilisation, but leaves the pigment fully oxygenated at the surface of the tissue.[156]

The oxyhaemoglobins of many nematodes (*Camallanus, Eustrongylides, Heterakis, Nippostrongylus, Haemonchus, Nematodirus, Ascaris* (body wall) and *Enoplus brevis*) become deoxygenated under virtually anaerobic conditions and re-oxygenate when oxygen is again available.[83,144]

The tendency of nematode haemoglobins to partially deoxygenate at relatively low oxygen tensions is an appropriate effect for a pigment that facilitates oxygen diffusion under poorly oxygenated conditions. This supplements oxygen supply when the diffusion of oxygen is inadequate for a normal level of aerobic metabolism. The haemoglobin in the pharynx of *E. brevis* is partly unloaded at low oxygen tensions (figure 4.9). Moreover, increased levels of activity at higher oxygen tensions cause deoxygenation of the pigment (figure 4.10). The haemoglobin in the pharynx of *E. brevis* may supply oxygen during periods of activity associated with the ingestion of food and it may enable the species to feed even when there is little oxygen available in its mud habitat.[157]

Unfortunately it is not certain that nematode haemoglobins do facilitate oxygen diffusion. The body wall haemoglobin of *Ascaris* does not facilitate oxygen diffusion *in vitro,* apparently because it does not have the rapid rate of dissociation from oxygen necessary for this effect. It has been suggested that either *Ascaris* haemoglobin is not typical of all nematodes or that rates of dissociation measured for these pigments *in vitro* may be lower than occur *in vivo.*[144]

Nematode haemoglobins that combine reversibly with oxygen may act as oxygen stores. The haemoglobin could release oxygen at low oxygen tensions in the environment and thus reduce the extent of changes in oxygen supply to a tissue that would otherwise be caused by fluctuations in activity of the tissue or of its oxygen supply. It seems probable that this will prove to be a secondary function in those nematode haemoglobins that are able to facilitate oxygen diffusion.[144,157]

4.5 Pharmacology

Chemicals that are used to remove animal-parasitic nematodes from their host are called anthelmintics, whereas chemicals that are used to destroy plant-

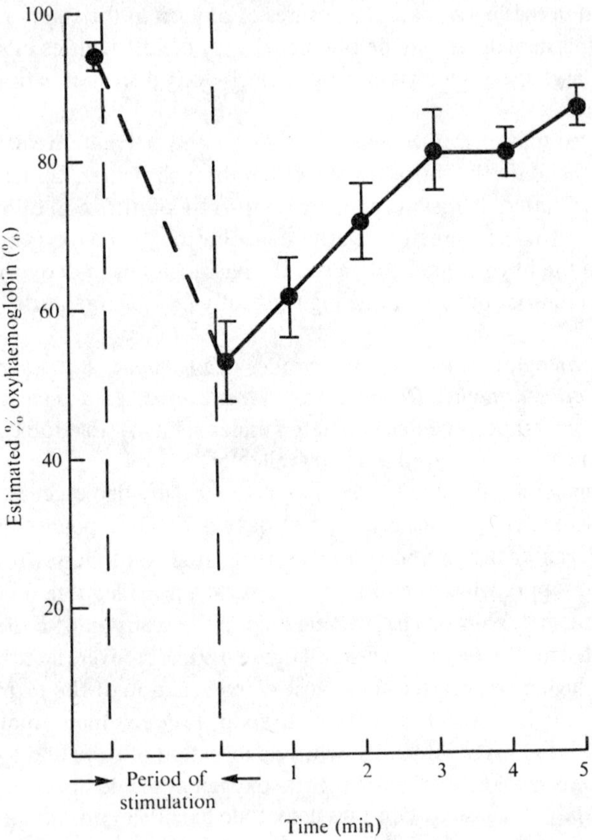

Figure 4.10 Estimated percentage of oxyhaemoglobin in the pharynx of the marine nematode *Enoplus brevis*, after stimulation. (After Atkinson.[157])

parasitic nematodes are called nematicides. Most nematicides are phytotoxic and so cannot be used to treat plant tissues that have become attacked by plant-parasitic nematodes. Several compounds are used as soil fumigants to destroy nematodes in the soil, but seem to act as general cytoplasmic poisons and are not specific for nematodes, for example methyl bromide, ethylene dibromide and dichloropropane-dichloropropane (DD). DBCP (1,2-dibromo-3-chloropropane), however, is a non-phytotoxic aklyl halide, which has little or no phytotoxic effect and is used to treat infected plants, but little is known about its mode of action.

A number of anthelmintics affect carbohydrate metabolism in nematodes and yet have little or no effect on the host. This indicates that they either interfere with metabolic pathways in the nematode that are either not present or can be by-passed by the host, or the host can alter the structure of the drug and so render it harmless to its own tissues.[158,159]

4.5.1 Glucose Uptake and Glycolysis

The anthelmintic dithiazanine interferes with the uptake and transport of glucose in adult *Trichuris vulpis* at concentrations that do not affect the mobility of this parasite. As a result of this inability to use exogenous glucose, the endogenous stores of carbohydrate are utilised, there is a decrease in the generation of ATP, and eventually this causes the death of the nematode.[158,159] The uptake of glucose by the microfilaria of *Dirofilaria immitis* is also inhibited by dithiazanine. This drug is toxic to some free-living nematodes, but its mode of action in these nematodes is not known. Similarly, low concentrations of mebendazole affect the uptake of glucose by adult *Ascaris suum, Syphacia muris* and *Syngamus trachea.* This results in a marked decrease in their endogenous glycogen but does not affect motility, so the worms continue to move and use up energy. The drug also inhibits the uptake of fructose by *Ascaris.*[160] Several benzimidazoles which possess anthelmintic activity also deplete the glycogen stores of nematodes. Thus parbendazole and cambendazole deplete the glycogen content of *Syphacia,* but thiabendazole, which is also a potent anthelmintic, only slightly affects the glycogen content of this nematode. It is interesting to note that mebendazole is much more active than cambendazole and parbendazole in stimulating the utilisation of glycogen by *Syphacia,* and its anthelmintic activity is several times greater than the other two drugs.[160] While there is no doubt that these anthelmintics do inhibit the uptake of glucose by these nematodes it is probable that some of the drugs exert an indirect effect, as they have been shown to interfere with the production of ATP. Dithiazanine, however, must have a direct effect on the uptake of glucose and not on glycolysis, as it allows motility of the nematodes and a decrease in the amount of glycogen in the tissues, that is, production of energy continues in the presence of the drug.

Some anthelmintics may interfere with steps in the Embden–Meyerhof pathway, for example, cyanine-863 inhibits the aldolase of *Litomosoides carinii* and a benzimadazole derivative partially inhibits the aldolase of *Ascaridia galli.*[159,161] Several trivalent, antimonial compounds interfere with glycolysis in schistosomes (Platyhelminthes) and stibophen (a well-known antischistosome drug) affects glycolysis in the microfilaria of *Dirofilaria immitis.*[159]

4.5.2 TCA cycle, Succinate Formation and Phosphorylation

The conversion of PEP to succinate via oxaloacetate, malate and fumarate appears to be an important source of energy in several animal-parasitic nematodes (section 4.1.8) and a number of important anthelmintics interfere with this pathway.[124,159,160]

Tetramisole, which is effective against a broad spectrum of animal-parasitic nematodes, inhibits fumarate reductase in adult *Ascaridia galli,* in adult *Ascaris* and in all stages of *Haemonchus contortus.* This drug also brings about contraction, followed by relaxation and irreversible paralysis, of whole adult *Ascaris.*

The drug first brings about activation of the muscles in the body wall, resulting in a breakdown of the myosin-bound ATP to ADP. Rephosphorylation of this ADP is inhibited because the drug inhibits fumarate reductase in the mitochondria of the muscle and consequently ATP becomes exhausted. It would appear that the fumarate reductase pathway in nematodes is especially sensitive to tetramisole as it does not inhibit the equivalent system in cestodes, nor does it affect the succinate dehydrogenase of the host animals.[159,160] Thiabendazole and cambendazole inhibit fumarate reductase in adult *Haemonchus contortus.*[162] The inhibitory effect of thiabendazole is partly relieved by high concentrations of cysteine, which suggests that the drug may be reacting with a sulphydryl group involved in the fumarate reductase mechanism. A strain of *Haemonchus* that is resistant to thiabendazole has a fumarate reductase that is insensitive to the drug at concentrations up to 1.0 mM and there is some cross-resistance to other benzimadazoles.[162]

Some of the classical uncouplers of phosphorylation in aerobic tissues have been shown to interfere with phosphorylation in *Ascaris.* This indicates that the phosphorylations in this nematode are linked to electron transport, but the final electron acceptor appears to be fumarate rather than oxygen. The anthelminitics dichlorophen, desaspidin and niclosamide, which are effective against cestodes but not against *Ascaris,* inhibit the mitochondrial incorporation of ^{32}Pi into ATP in homogenates of this nematode. The reason these anticestode drugs are not effective against *Ascaris in vivo* may be because they are unable to penetrate the tissues or the mitochondria of the intact nematode. Dithiazanine, which affects the uptake of glucose by some nematodes (section 4.4.1), also inhibits the mitochondrial incorporation of ^{32}Pi into organic phosphate and thus may also function by inhibiting electron-transport-associated phosphorylation.[124,159,160]

Ethylene dibromide kills soil nematodes slowly at low concentrations and they die in a relaxed state; high concentrations kill them quickly and they die contorted. It is probable that at low concentrations the alkyl halide rapidly oxidises the iron centre in the respiratory chain, producing narcotisation and a slow death, whereas at high concentrations alkylation of reactive sites of proteins and co-enzymes leads to a more rapid death.[163]

The amount of succinate produced by adult *Ascaris* falls sharply when the nematode is paralysed by piperazine but it appears that this is a secondary effect brought about by the paralysing action of the drug on the muscle (see section 7.3), the paralysed muscle requiring much less ATP and, therefore, producing much less succinate via the fumarate reductase pathway.[158,159,160]

See section 7.3 for an account of the effect of drugs on the nervous system and muscles of nematodes.

5 Osmotic and Ionic Regulation: Excretion

5.1 Osmotic Regulation

5.1.1 Introduction

Virtually all animal cells maintain an intracellular content of solutes which differs from the medium surrounding them. More complex animals usually possess a second extra-cellular compartment of body fluid which partially isolates the cells from the external environment. The body fluids may have a similar osmotic pressure to the surroundings (isosmotic) but frequently contain either more solutes (hyperosmotic) or sometimes less solutes (hypo-osmotic) than the external medium. A number of animals can tolerate quite large variations in their internal osmotic pressure following a change in the external osmolarity (osmoconformers). Many animals that are able to tolerate wider environmental fluctuations are capable of regulating the water or ionic content of their body fluids (osmoregulators). There is no rigid division between these responses to osmotic stress; some animals are osmoconformers in one concentration range but regulate in another.[164] Nematodes are likely to possess some of these abilities, particularly those species that normally experience variations in the osmotic pressure of their aquatic habitats.

5.1.2 Volume Regulation in Nematodes

The ability to maintain a constant volume in conditions of osmotic stress occurs in many animals and has particular significance in nematodes which depend on a hydrostatic skeleton and muscle tone for locomotion. Such changes have frequently been studied in nematodes but usually volume changes have been inferred from variations in length. Comparison of species is not always possible with measurements of this kind for nematodes are not of constant shape, and a volume change does not necessarily influence length and radius equally in all species.[139]

5.1.2(a) Marine Nematodes

The sea contains a considerably higher concentration of dissolved ions than most other aquatic environments. Many marine invertebrates, including those nematodes investigated, are isosmotic to this environment.[139,165] Many have limited ability to tolerate changes in the osmolarity of sea water. Changes in length of the marine nematode *Enoplus communis* indicate that its volume is increased in dilute sea water and is reduced when the animal is placed in a hyperosmotic medium of partially evaporated sea water. This enoplid is usually covered by sea water but *Enoplus brevis* is a mud-dwelling species of this genus which is not exclusively marine. It cannot be considered as euryhaline in distribution but does occur in salt marshes or estuaries where there are tidal fluctuations in salinity. This nematode can restore its original body volume in hypotonic sea water and also shows a limited ability to do this in hypertonic solutions (figure 5.1).[139] The differences between the two species of *Enoplus* are a clear indication that the ability of a nematode to regulate its volume is an adaptation to the particular requirements of its habitat.

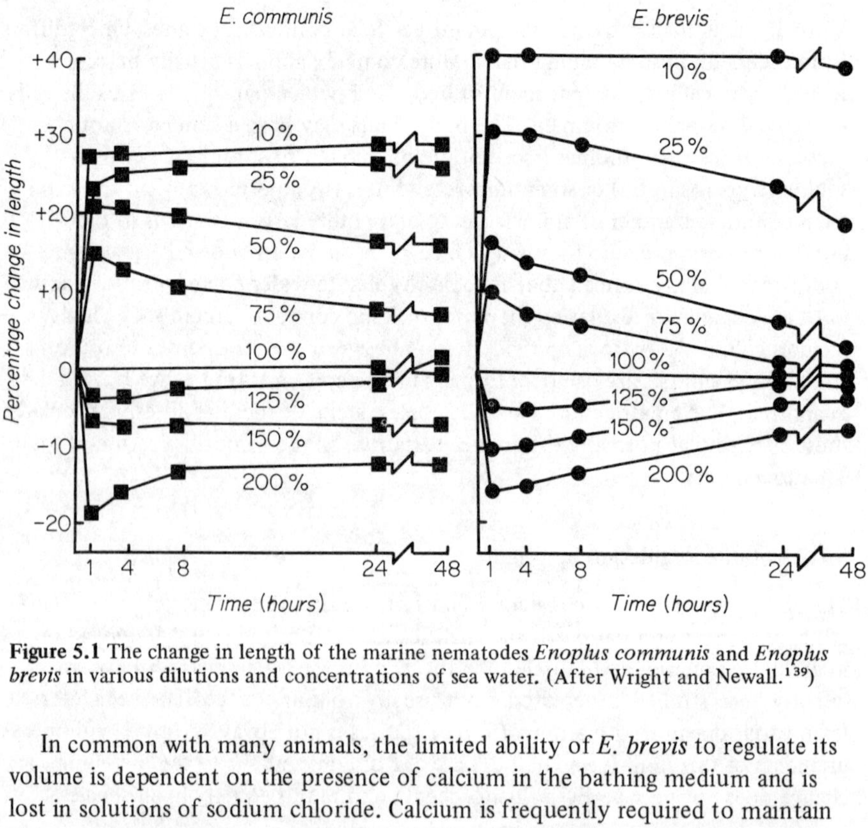

Figure 5.1 The change in length of the marine nematodes *Enoplus communis* and *Enoplus brevis* in various dilutions and concentrations of sea water. (After Wright and Newall.[139])

In common with many animals, the limited ability of *E. brevis* to regulate its volume is dependent on the presence of calcium in the bathing medium and is lost in solutions of sodium chloride. Calcium is frequently required to maintain the normal permeability characteristics of membranes and intercellular cements. Unfortunately, nematode physiologists have frequently used unbalanced salt solutions that are of questionable value for animals that usually experience a wide range of ions in their environment.[139]

There is normally a sharp distinction between marine and fresh-water groups of nematodes. Most estuarine forms are marine species that have extended their range into diluted sea water. The remainder belong to distinct groups of fresh water origin and are mainly found in low salinities. There are relatively few euryhaline nematodes but some may exist. *Aphelenchoides parienturis* and *Panagrolaimus rigidus* can tolerate salinity of forty parts per thousand.[166] *Rhabditis marina* can be cultured in salinities ranging from virtually 0 to 230 per cent sea water, but its intertidal habitat is normally about 80 per cent sea water.[92] A thorough study of volume and ionic regulation by a euryhaline nematode would be useful in developing our knowledge of the processes used by nematodes to withstand extreme osmotic stress.

5.1.2(b) Soil- and Fresh-water Nematodes

The quantity of solutes dissolved in soil water may vary but they always exert a much lower osmotic pressure than sea water. Fresh water can have a very low salt content but little is known of the osmotic relations of fresh-water nematodes.

Nematodes inhabiting soils experience fluctuations in the osmotic pressure of their surroundings. Rainfall adds water and tends to reduce the osmotic pressure. Evaporation tends to dry the soil and increases its osmolarity. However, the fluctuations in water content of a soil are usually less than a few fold except in the surface layer of the soil and in arid areas. Such changes are not beyond the limits tolerated by an efficient osmoregulator.

Unfortunately, the water relations of nematodes are complicated by the nature of soils. Water in a partially dry soil is mainly held in small pores and in films in the spaces between the larger particles (see section 8.2.3). The magnitude of the force resisting further water loss from a soil is known as the matric potential. Its value is increased if evaporation, drainage and transpiration by plants succeed in reducing the water content.[167] There is a distinction between the force that a nematode must overcome to take water from the soil and the osmotic pressure exerted by dissolved solutes in free water. A decrease in water content will immobilise a nematode long before the matric potential results in the wilting of plants. Clearly, the activity of nematodes is susceptible to dry conditions but many species are adapted to survive at such times in an inactive state. The role of osmoregulation in dry soils is not very clear; the relevance of homeostasis is somewhat reduced by conditions unsuited for activity. Volume regulation may be of value in ensuring that osmotic stress does not impair activity in conditions that are otherwise suitable for locomotion, but survival mechanisms such as cryptobiosis would seem to predominate in soils of a very high matric potential.

Rhabditis terrestris is an example of a nematode that is liable to experience fluctuations in the osmotic pressure of its surroundings. A resting juvenile occurs in the earthworm and feeds on decaying tissues and accompanying bacteria after the death of the animal. A few generations may occur before decomposition is complete and juveniles disperse through the soil to other earthworms. The nematode must adjust osmotically to life in the soil solution, the earthworm and its decomposing cadaver. Changes in length suggest that the animal regulates its volume in solutions hypo-osmotic to its culture medium including distilled water.

The animal also has a limited ability to reduce initial shrinkage in hypertonic solutions. Furthermore, the nematode is isotonic to a normally hyperosmotic medium after it has lived for a few generations in a medium containing a balanced salt solution with a raised sodium chloride content. It seems that the animal can adjust to rapid changes in osmolarity and is probably able to adjust to the more gradual changes typical of its natural environment without undue body distortion.[168]

5.1.2(c) Plant-parasitic Nematodes

Heterodera is one example of an endoparasitic nematode that has a juvenile stage capable of withstanding both fluctuations in the osmotic pressure in the soil and also a change in environment associated with the penetration of a plant. Second-stage juveniles of *Heterodera rostochiensis* change volume considerably in response to osmotic stress, but adjustment is indicated by relatively small changes in length. The animal is capable of regulating its volume in hypotonic media, but not those hypertonic to its tissues (figure 5.2).[139] Little is known of the effects

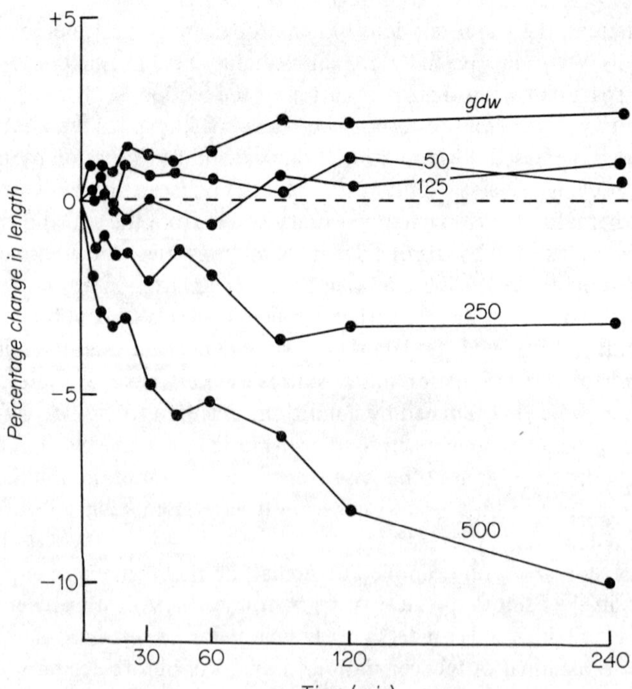

Figure 5.2 The change in length of second-stage juveniles of *Heterodera rostochiensis* in glass distilled water (gdw) and artificial tap water supplemented with 50, 125, 250 and 500 mM NaCl. (After Wright and Newall.[139])

of osmotic stress in plant tissues on endoparasitic nematodes. Such changes must occur, particularly in plants parasitised by *Heterodera* and *Meloidogyne*, for heavy infection by these nematodes can cause plants to wilt.

5.1.2(d) Animal-parasitic Nematodes

Many animal-parasitic nematodes are approximately isotonic with the body fluids of their hosts. Their osmotic relations will be stabilised by the regulatory processes of the host but may still be liable to some fluctuations. *Hammerschmidiella,* from the hind-gut of the cockroach, experiences marked changes in osmotic pressure during moulting of the insect. The nematodes escape from the peritrophic membrane, which is cast during ecdysis, into the posterior part of the hind-gut and subsequently move forward to their more normal position. The nematode is able to withstand the increase in osmotic pressure associated with the moulting process. Changes in length suggest that it is able to regulate its volume in hypertonic solutions of sodium chloride (figure 5.3). A similar but slow response also occurs in sea water diluted to give a more balanced hyperosmotic solution.[169] The animal shows little ability to restore volume in hypo-osmotic media; perhaps such changes do not occur in its habitat.

Angusticaecum, from the colon of the tortoise, may also experience changes in osmotic pressure related to the variations in water content within the gut of

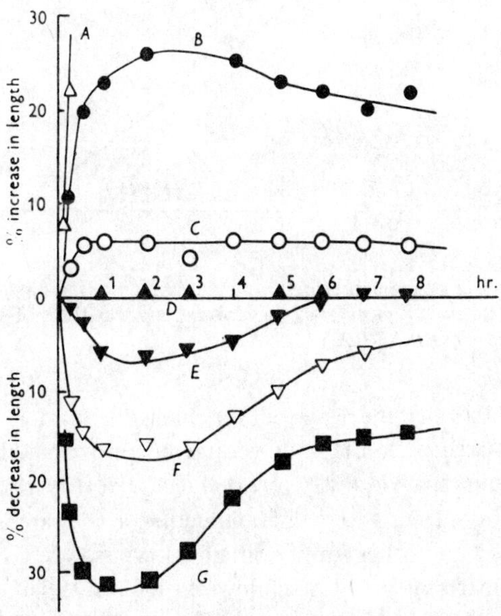

Figure 5.3 The percentage change in length of females of *Hammerschmidtiella diesingi* in distilled water and in various concentrations of sodium chloride. *A.* Distilled water; *B.* 0.05 M NaCl; *C.* 0.08 M NaCl; *D.* 0.15 M NaCl; *E.* 0.2 M NaCl; *F.* 0.3 M NaCl; *G.* 0.5 M NaCl. (After Lee.[169])

its host. Limited evidence suggests that the animal is able to regulate its internal osmotic pressure in tap water but it is an osmoconformer in hypertonic solutions.[216]

It is difficult to measure the osmotic pressure of gut contents but *Aspiculuris tetraptera* is isotonic with 35 per cent sea water. It can regulate its volume in a hypotonic solution of 20 per cent sea water and also in hyperosmotic media.[170] *Ascaris* does not regulate its volume in anisotonic solutions *in vitro* but this may reflect its deteriorating physiological condition *in vitro* rather than its normal response *in vivo.*[139,171]

Juveniles of some animal parasites hatch in the faeces of a parasitised animal and migrate through soil before invading a host. They must withstand changes in osmotic pressure associated with these changes in habitat. Both *Nippostrongylus* and *Ancylostoma* possess an ampulla associated with their excretory system and this ampulla shows a pulsation frequency inversely proportional to the osmotic pressure of their surroundings (figure 5.4). There is an appreciable flow of fluid

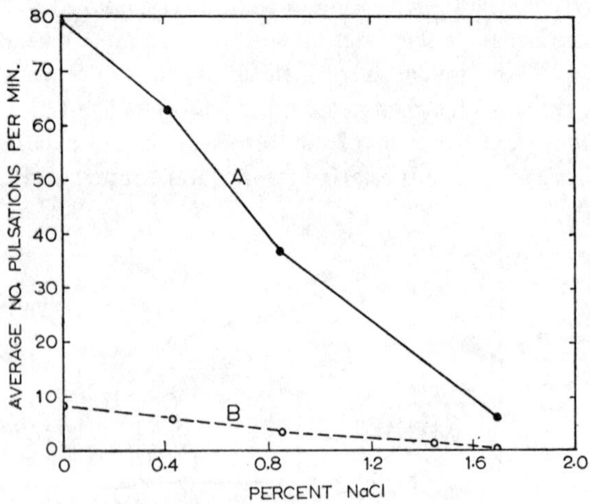

Figure 5.4 Pulsation rates of the excretory ampulla of A, *Nippostrongylus brasiliensis* larvae and B, *Ancylostoma caninum* larvae in various concentrations of sodium chloride. [After Weinstein, *Exp. Parasit.*, (1952).]

through their ampulla when the nematodes are in distilled water. The amount of fluid expelled is equivalent to the body volume in approximately eleven hours and seventy-five hours for *Nippostrongylus* and *Ancylostoma* respectively.[188]

Our present knowledge suggests that marine nematodes are isosmotic with their surroundings. Many other forms contain a lower solute content but are hyperosmotic to environments of much lower osmolarity than sea water. The ability to regulate volume may be important for locomotion but would not seem to be a general ability shared by all species in both hypo-osmotic and hyper-osmotic media. As in other animals, the well-developed ability of some nematodes to regulate their volume is an adaptation for life in particular habitats.

5.1.3 Permeability to Water

Osmosis tends to degrade the internal organisation of animals but some permeability to water is essential for assimilation and excretion. However reduced permeability lowers the flux of water between the animal and a dissimilar osmotic pressure of the environment. It is therefore hardly surprising that a reduction in general permeability to water in animals is frequently associated with adverse osmotic conditions in their habitats.[164] In some cases much of the exchange of water and solute with the environment is restricted to specific regions of the animal.

Marine invertebrates are usually isosmotic with sea water and are freely permeable to water. The marine nematode *Enoplus* is no exception and has a relatively high over-all permeability to water of 4.4 $\mu^3 H_2O/\mu^2$ body surface/s.[139] The soil-inhabiting form *Aphelenchus avenae* has a much lower water permeability of 0.0075 $\mu^3 H_2O/\mu^2/s$;[172] this value is similar to that of many fresh-water animals. Nevertheless, changes in tissue hydration will be rapid even in these species because of the high surface/volume ratio of small nematodes. The adaptive significance of this reduced permeability is that it lessens the intensity of the osmoregulatory processes necessary to maintain a normal level of hydration in media of differing osmolarity.

Exchange of water probably occurs across all the body surfaces of a nematode but some regions may be more permeable than others. In hypotonic solutions the intestine of *Rhabditis* pumps water out of the animal, and *Ascaris* and *Enoplus* also pass water out through the anus.[139,168,173] It would seem that the permeability of the gut in these nematodes is sufficient to eliminate a large proportion of water diffusing into the animal. The mechanisms involved in fluid transport in the gut of *Ascaris* are described in sections 3.5 and 5.2.3. The pulsating ampulla in the excretory system of the juveniles of *Nippostrongylus* and *Ancylostoma* probably has a similar role. Assuming that they have a similar order of permeability to that in *Heterodera* and *Aphelenchus*, the infective juveniles of these animal-parasitic nematodes can expel sufficient water through their excretory system to account for the total influx of water from a hypotonic medium.[139]

5.1.4 Desiccation Survival

Nematodes require an aquatic medium for activity but certain species can survive periods without water in a dormant state. The water content of the nematode does fall during this time but their survival is a specialised ability and is not an aspect of osmoregulation. Nematodes survive desiccation in a cryptobiotic state with their metabolism virtually suspended, and some species may respond in a similar manner to oxygen lack, freezing, or extreme osmotic stress.

Many of the relatively few species of nematode capable of withstanding desiccation are parasites of the aerial parts of plants, such as leaves, stems or inflorescences (*Ditylenchus dipsaci, Anguina tritici*) and are more liable to experience dry conditions than those species which remain in soil. Little is known of the mechanisms which enable nematodes to maintain their viability and structural integrity when they are desiccated. However, it does seem that individuals that can survive these conditions are able to retard their rate of drying when compared with other forms without this ability. *Panagrellus* and second-stage juveniles of *Heterodera rostochiensis* dry more quickly than fourth-stage juveniles of *Ditylenchus dipsaci*, which is more able to withstand air-drying. The second-stage juvenile of *Heterodera rostochiensis* can also withstand desiccation while within its egg shell, particularly if the eggs are not removed from the cyst formed by the dead body wall of the female. Both the cyst wall and the egg shell tend to reduce the rate of water loss from the juveniles in dry conditions. Similarly, infective juveniles of *Haemonchus contortus* lose water less readily in the ensheathed form, which is capable of withstanding air-drying on grassland, than when exsheathed after ingestion by their host (figure 5.5). It seems that both the sheath of *Haemonchus* and the egg shell of *Heterodera rostochiensis* are freely permeable to water when wet but become more effective barriers to water loss as they dry. It is a retardation in the rate of drying that seems to be important in desiccation survival and the final water content of tolerant forms is similar to that of species that are unable to withstand drying.[174] Unfortunately, the significance of the reduction in the rate of drying will only be fully appreciated when the mechanisms of desiccation survival in nematodes are more completely understood.

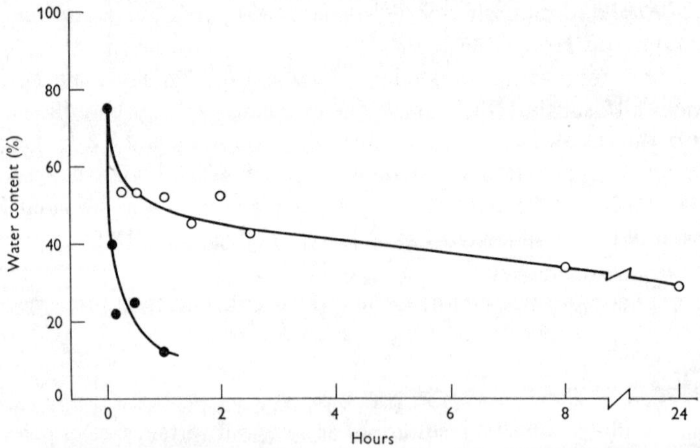

Figure 5.5 Water loss of exsheathed (•) and ensheathed (○) juveniles of *Haemonchus contortus.* (After Ellenby.[174])

5.2 Ionic Regulation

5.2.1 Introduction

The limited ability of some nematodes to restore body volume after a change in the external osmotic medium indicates the ability to osmoregulate. It does not necessarily follow that these nematodes are capable of ionic regulation. Some animals respond to osmotic stress by altering their intracellular osmotic pressure with solutes other than ions. Changes in free amino acids are particularly important in the muscles of some animals but have usually been recorded in adjustment to hypotonic solutions (*Carcinus, Mytilus, Nereis diversicolor*).[164] Such mechanisms will not, however prevent the loss of certain ions along a concentration gradient to the external medium.

Unfortunately, there has been no attempt to assess the extent of intracellular regulatory processes in nematodes. The release of non-electrolytes into the cytoplasm is considered a primitive but important ability in many euryhaline invertebrates. It seems possible therefore that this mechanism plays a part in the volume regulation of nematodes. However it has to be admitted that less is known of the extent of ionic regulation in nematodes than in most other invertebrates. This is mainly due to difficulties associated with the use of standard techniques with animals of the small size typical of most nematodes.

5.2.2 Ionic Regulation in Nematodes

The active uptake or elimination of ions has not been demonstrated in any species, but sampling of the pseudocoelomic fluid of *Ascaris* has shown that this nematode can maintain the ions in this fluid at different concentrations from those of its surroundings. Transfer to hypotonic 30 per cent sea water causes a slight increase in volume but the ions are maintained in the pseudocoelomic fluid at a different concentration from that in the bathing medium. On this basis it has been suggested that the animal can maintain constant chloride, magnesium, calcium and potassium contents of its pseudocoelomic fluid in slightly hypotonic media.[171] *In vivo* the animal probably maintains a slightly dissimilar ionic content from that of the isosmotic or slightly hypo-osmotic gut fluid.[89]

Even less is known of ionic regulation in smaller nematodes. The work necessarily involves the analysis of large numbers of nematodes or at best whole animals. It is therefore not possible to be certain of the concentration of an ion in a particular compartment of the individual. Much of the sodium might be expected to be extra-cellular but the proportion measured actually outside the animal and within the gut lumen of the nematode cannot readily be estimated. Accepting the limitations of the available data, some evidence is available for ionic regulation by small nematodes. A comparative study has shown that the marine nematode *E. communis* loses sodium to hypotonic dilutions of sea water more readily than does the closely related *E. brevis* from a more estuarine environment. This difference may not be due to passive processes for both gain sodium equally in hypertonic solutions made from partially evaporated sea water[139] (figure 5.6).

Panagrellus redivivus is able to limit the increase in its potassium concentration in a medium with a raised content of this ion but it does not regulate its sodium content in either glass-distilled water or in a hypertonic saline.[175] The second-stage juvenile of *Heterodera rostochiensis* is also unable to regulate sodium in glass-distilled water but it does limit the increase in total sodium in the body when placed in tap water supplemented with sodium chloride.[139] This suggests that *Heterodera* does have a limited ability to regulate its extra-cellular sodium content.

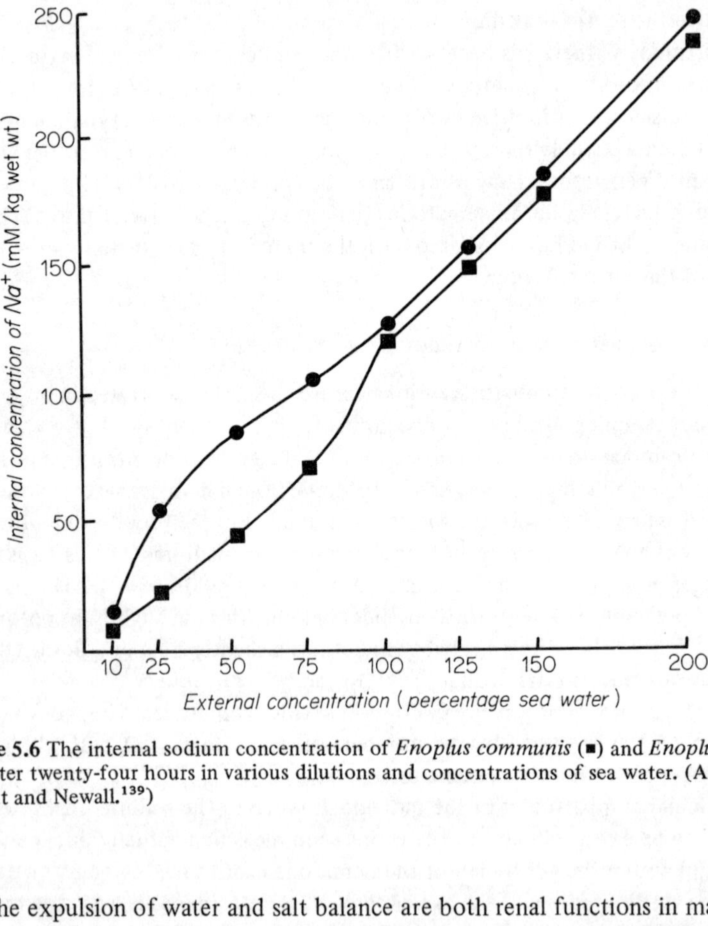

Figure 5.6 The internal sodium concentration of *Enoplus communis* (■) and *Enoplus brevis* (●) after twenty-four hours in various dilutions and concentrations of sea water. (After Wright and Newall.[139])

The expulsion of water and salt balance are both renal functions in many animals but there is only limited evidence for the latter process in the nematode excretory system. Early work suggested that ionic regulation of *Ascaris* involved structures in the body wall.[171] It has been suggested more recently that transfer of *Ascaris* from pig intestinal fluid to a hypotonic solution of 30 per cent sea water reverses the normal potassium/sodium ratio of the fluid within the excretory system. If the animal is then replaced in a fluid with a higher potassium content the ratio in the excretory system is restored in favour of a higher potassium excretion.[176] Unfortunately no values are available at present for the quantity of fluid and ions expelled in this way.

Clearly, there is limited evidence for ionic regulation in nematodes. It would be surprising if the cells of nematodes proved unable to actively eliminate sodium as this is a virtually universal ability in animals and is related to the maintenance of a high intracellular content of potassium. Active transport of sodium is also an essential factor in the conduction of nerve impulses (see section 7.2.3). There is some evidence that second-stage juveniles of *Heterodera rostochiensis* regulate sodium to a limited extent in hypertonic media and many other soil-inhabiting nematodes may have a similar ability. The reduced loss of sodium from *Enoplus brevis* in hypotonic solutions compared with the more stenohaline *Enoplus communis,* is slight evidence for a limited ability of this nematode to regulate sodium in hypotonic solutions.[139]

Other nematodes may have better-developed powers of ionic regulation. *Ascaris* is able to maintain the potassium, magnesium, calcium and chloride contents of its pseudocoelomic fluid in spite of environmental fluctuations. However, active uptake or excretion of an ion has never been adequately demonstrated in this nematode and the structures concerned in salt balance have not been unequivocally identified.

5.2.3 Permeability to Solutes

Animals are not equally permeable to all solutes, and lipid components of cell membranes are mainly responsible for this selectivity. Lipophilic substances tend to pass more readily into cells than those which are water soluble. The penetration of organic molecules into many nematodes may be a diffusion-controlled process in which neutral non-polar lipophilic molecules penetrate the nematode with least difficulty. The efficacy of many organophosphates, alkyl halides and oxime carbamates as nematicides is partly due to their lipophilic nature. They readily penetrate the individual and may also tend to accumulate within the nematode because their solubility at lipophilic sites is significantly greater than their solubility in the aqueous medium surrounding the animal.[163,177,214]

The cuticle and the underlying membrane of the hypodermis of most nematodes are slightly permeable to a few solutes in aqueous solutions.[178] Some small ions (for example sodium and ammonium ions and protons) and solutes such as urea may pass out through the cuticle of *Ascaris*, but this animal cannot absorb physiologically significant quantities of glucose from the environment by this route.[179] This may not be true for all nematodes. *Bradynema* absorbs nutrients from the haemocoele of its host through its modified body surface and lacks a functional gut.[28] However the intestine is undoubtedly a most important region for the passage of solutes in the majority of nematodes. The digestion of nutrients is considered elsewhere (section 3.3), but the mechanisms involved in solute movement through membranes are considered in this section. The general mechanisms involved are probably characteristic of the movement of solutes through most of the membranes which surround the cells of nematodes.

There are a number of different ways by which materials that are hydrophilic can cross membranes and their relative physiological significance depends on the solute considered. In the absence of other factors, all solutes will diffuse through a permeable membrane down their concentration gradient. *Ascaris* has been used for studies on uptake of nutrients because its large size facilitates experimentation. Its intestine consists of an epithelium one cell in width. Recent work has shown that it transports fluid through the membrane bordering the mucosa. The fluid passes out through the lateral membranes of the cells into intercellular spaces which connect with the pseudocoelom (see figure 3.9). Similar systems occur in many animal epithelia and the particular type found in *Ascaris* is frequently correlated with the movement of water as the result of osmosis and diffusion following the active transport of sodium. The transported fluid in the type of system found in *Ascaris* probably has a similar osmolarity to that of the gut and pseudocoelomic fluids but can involve relative changes in solute concentrations. The fluid flow is closely related to glucose transport and the mechanism probably represents an important pathway for the uptake of glucose by *Ascaris in vivo.*[89]

The membranes of nearly all animal cells actively exude sodium and maintain this ion at a low intracellular level in favour of a high potassium content. There is some evidence for the presence of a sodium pump in nematodes and the iso-osmotic transport of water suggests that active transport of sodium does occur in the intestine of *Ascaris*. The active transport of monosaccharides and amino acids against their concentration gradient is known to be by similar mechanisms and frequently requires the presence of sodium. The ovary of *Ascaris* is able to accumulate amino acids from the pseudocoelom by active transport but the mechanism may not be present in the muscle system.[180] Active transport of calcium, magnesium, potassium and phosphate by cell membranes is also known to occur in some animals but these processes have not been studied in nematodes.

Differences in concentration of ions across a membrane does not necessarily imply active transport. In particular chloride may diffuse down an electrical gradient across a membrane resulting in a dissimilar concentration of the ion in two compartments. This mechanism could be involved in the difference in chloride concentration between the pseudocoelom and the exterior of *Ascaris.*[171] Ammonium ions are produced in excess by ammonotelic nematodes. This ion and some others may be exchanged across membranes for other cations in the environment.

Pinocytosis represents a further method for the movement of solutes across membranes but there is little evidence for this process in nematodes at present. Finally, it is possible that solute transport might be increased by the bulk flow of water associated with a filtration pressure. This occurs in the mammalian glomerulus but the high internal pressure of many nematodes has not been shown to be of physiological significance in enhancing solute flow into either the excretory system or the gut lumen.

5.3 Excretion

5.3.1 Introduction

The nitrogen excreted by animals has two main origins. Amino acids from excess absorbed proteins cannot be stored and are deaminated. The nitrogen released is excreted but the carbon skeleton is used as an energy source in carbohydrate metabolism. Secondly, a fraction of assimilated amino acids undergo deamination during the normal turnover of body proteins. The amino acids undergo catabolism in a similar way in both cases. First each undergoes a series of transaminations in the mitochondria or cytoplasm until the amino groups have collected in a single form; in most animals this is usually glutamic acid. This amino acid is then deaminated by a specific glutamate dehydrogenase. This enzyme therefore plays a major role in nitrogeneous excretion, particularly in animals that excrete appreciable quanties of urea.

5.3.2 Products of Nitrogen Catabolism in Nematodes

These processes leading to the excretion of nitrogen probably occur in nematodes.[139] Transaminases, which produce metabolites for the tricarboxylic acid cycle, have been demonstrated in *Caenorhabditis.*[181] The ovaries of *Ascaris* contain alanine and aspartate deaminase, both of which catalyse the transfer of amino groups to form glutamic acid.[182] They also possess a weak glutamate dehydrogenase but the muscle of *Ascaris* has a higher activity of this enzyme, which may indicate a greater potential rate of deamination in this tissue.[183]

Glutamate dehydrogenase has also been detected in *Panagrellus* and *Haemonchus* and another, less specific deaminase of uncertain function, has been found in *Ascaridia* and *Nematodirus.*[139]

Most animals excrete the nitrogen derived from exogenous and endogenous origins in one predominate form, usually as ammonia (ammonotelism), urea (ureotelism) or uric acid (uricotelism). Ammonia and its ion are extremely soluble and diffusable in water but are extremely toxic to most animals. It is therefore typically the main excretory product of aquatic invertebrates, which are unlikely to accumulate appreciable concentrations of ammonium ions in their tissues. Most nematodes are small aquatic animals and it is not surprising that all species investigated are predominantly ammonotelic, excreting between 40 and 87 per cent of non-protein nitrogen as ammonia or its ion (table 5.1).

The toxicity of ammonia to tissues is due to a number of effects: it alters cell pH, it influences membrane permeability and it tends to reverse the glutamase dehydrogenase reaction. This last effect tends to suppress the tricarboxylic acid cycle by depletion of α-ketoglutarate (see figure 4.2).[164] It might be anticipated that production of ammonia will be reduced in environmental conditions that are liable to raise its tissue concentration. This does occur in *Panagrellus* and *Ascaris*, but it is uncertain if the toxic action of excess ammonia reduces the over-all level of metabolism or causes a more specific suppression of deamination.[139] An early report suggested that *Ascaris* is capable of reducing ammonia production in favour of increased urea excretion if placed in tubes limiting the medium available for ammonia excretion but this has not been substantiated (table 5.1).[184] Perhaps many animal-parasitic species rely on the efficient detoxification processes of their host to remove their excreted ammonia.

Other nitrogenous compounds have been detected in nematode tissues and in the medium surrounding the animals. Unfortunately some of these compounds may not be true excretory products but may have a range of other origins. Most animals eliminate some unabsorbed nitrogen from their diet in the faeces; amino acids may leak even from undamaged individuals and nitrogenous material may also be secreted for various functions outside the animal (section 5.3). Much of the urea apparently excreted by *Ascaris* may diffuse into the animal with its diet and is probably not a product of deamination by the nematode. However many nematodes probably excrete a little urea. *Panagrellus* has all the necessary enzymes for the ornithine–arginine cycle but the enzyme activities are low and it is probably not quantitively significant in nitrogen excretion. The pathway may be required for the biosynthesis of arginine, as in micro-organisms.[185]

Nucleic acids are a quantitatively minor, but important, part of nitrogen metabolism in animals. They are normally degraded to mononucleotides and eventually yield purines or pyrimidine bases by enzymatic hydrolysis. If excreted they are degraded enzymatically to uric acid, allantoin, urea and finally to ammonia in a few animals. Small quantities of any of these compounds recorded in nematodes might be derived from this aspect of nitrogen metabolism. *Panagrellus* possesses some activity of the enzymes required to form at least allantoin[185] but it does not possess the detectable levels of urease that enable *Ascaridia, Ascaris, Nematodirus* and *Haemonchus* to convert the urea formed to ammonia.[186]

Although peptides and free amino acids have been reported as excretory products of nematodes[186] it is difficult to demonstrate that these products represent true excretory products rather than having one of the other origins previously discussed. Several enzymes, including amino peptidase, are released from the excretory pore of *Nippostrongylus* and may have an exodigestive role in this parasite. This species also secretes acetylcholinesterase, which may act as a biochemical holdfast[11] (section 5.3). Peptides and free amino acids may also have a secretory function in some nematodes and cannot be considered as excretory products in these cases. It has been suggested that the excretion of free amino acids is associated with a reduction in ammonia toxicity.[181,186] This is a possibility, but their excretion represents a waste of potential energy for it seems likely that the carbon skeleton could have entered carbohydrate metabolism after trans-deamination. Apparent excretion of amino acids in *Panagrellus* probably represents the release of non-assimilated material for their rate of loss falls more rapidly in starved animals than does the production of ammonia or urea. Also the loss of amino acids is enhanced in conditions, known to be detrimental to the worms, that probably increase leakage of body contents. This suggests that

Table 5.1 Nitrogenous Excretion in Nematodes[139,186,187]

NEMATODE	Expt. conditions	NH_3 N %		Urea N %		Total Nitrogen μMN/g/24 h	Body size (radius in mm)	Other nitrogenous products passed out by nematodes to media (% of total nitrogen)			
		a	*b*	*a*	*b*			Peptide	Amino acid	Amine	Uric Acid
Trichinella (juvenile)	Aerobic	42	33	0	0	200	0.01	21	28	7	0
Ditylenchus (mixed pop.)	Aerobic		39	0	0	10	0.01	+	28	+	0
Panagrellus (mixed pop.)	Aerobic	60		12		213	0.02				
Nematodirus (adults)	Aerobic	63	42	22	14	97	0.080	35	+	–	3
Nematodirus (adults)	Anaerobic	45	29	6	4	123	0.080	35	+	–	4
Ascaridia (adults)	Aerobic	66	56	9	12	25	0.5	15	+	–	0
Ascaridia (adults)	Anaerobic	70	59	9	15	26	0.5	15	+	–	0
Ascaris (adults)	Aerobic	87	69	9	7	28	2.51	21	+	–	0
Ascaris (adults)	Anaerobic	87	71	10	6	29	2.5	18	+	–	0
Ascaris (adults)	'U' tube	35	27	64	51	28	2.5	19	–	–	0
Ascaris (adults)	Aerobic	13 μM		3 μM		–	2.5				
Ascaris (adults)	'U' tube	4 μM		2 μM		–	2.5				

a Percentage of total non-protein nitrogen[187]
b Percentage of total nitrogen[186]

the amino acids do not represent a major form of excreted nitrogen in this nematode.[187]

Amines have also been detected in media surrounding the infective juveniles of some animal-parasitic nematodes (*Trichinella, Ascaris, Nippostrongylus*) and occur in the fluid that bathes juveniles of *Ascaris* enclosed within their egg shells. The physiological significance is unknown but some amines could be derived from the decarboxylation of amino acids. Further confirmation is required before peptides, amino acids and amines could be considered true products of quantitative significance in the normal nitrogeneous excretion of any nematode species.

The over-all rate of nitrogen metabolism of a nematode is likely to vary with a variety of endogeneous and environmental factors. The rate of metabolism decreases with increasing body size (section 4.4.2) and consequently the rate of nitrogen excretion may also vary in similar manner. The proportion of proteins in the diet will also affect nitrogen catabolism. A high protein intake tends to enhance the rate of transdeamination and increases the proportion of protein utilised in energy metabolism; starvation and a low-protein diet will have the opposite effect.[139]

5.3.3 Other Excretory Products

Carbon dioxide is a major end-product of energy metabolism that is excreted in all nematodes. The ambient concentration influences the oxygen consumption in *Ditylenchus dipsaci.*[147] This may be a response associated with host location and not a direct effect on metabolism. Carbon dioxide produced by nematode tissues does not readily have a toxic effect on the animal. The gas is very soluble and probably diffuses rapidly into most environments because of the favourable surface/volume ratio in nematodes. Carbonic anhydrase may be present in the tissues of a few nematodes that do experience toxic levels. The enzyme hydrates the gas to carbonic acid, which is mainly dissociated at physiological pH. The bicarbonate ion is present in the pseudocoelomic fluid of *Ascaris* and this nematode may eliminate some carbon dioxide in this form.

A few nematodes also excrete end-products of incomplete carbohydrate metabolism. Lactic acid is excreted by some (*Litomosoides, Dracunculus*) but only traces of it are excreted by many of the other nematodes investigated (section 4.1.5). Some animal-parasitic nematodes also release mixtures of fatty acids, which may be normal end-products of carbohydrate metabolism (section 4.1.8).

5.3.4 Excretion of Nitrogenous Waste

Most nematodes contain structures termed the excretory system but available evidence suggests that they are rarely the major route of nitrogenous excretion. Nematodes are usually ammonotelic and excrete ammonia in either an ionised (NH_4^+) or a non-ionised (NH_3) state. Ammonia is almost completely ionised at normal physiological pH *in vivo* but the general body surface of nematodes is probably relatively impermeable to ammonium ions. Many ammonotelic animals excrete much of this ammonium across specialised regions of the animal; this could occur through the gut wall in nematodes. Loss of ammonia will also be favoured by the large surface/volume ratio typical of smaller nematodes. The equilibrium between the two states may ensure a constant flux from the ionised form to replace the lost non-ionised molecules. It may prove that ammonia is the major pathway for nitrogen excretion in smaller nematodes. The loss of ammonia may also be substantial in larger animal-parasitic nematodes, for its outward diffusion is favoured by the relative acidity of the surroundings.[139]

The direction of diffusion of ammonia from a tissue may be influenced by its position in the animal. Muscle is responsible for much nitrogen catabolism in *Ascaris* and any free ammonia released is likely to diffuse out through the body wall. The intestine is also an important centre for deamination, but the ammonia from this source is likely to pass into the gut contents. If the ammonium ions are also excreted from the gut the pseudocoelom could play a role in their transport from the muscle.[139]

Most animals are permeable to urea and the compound passes freely into *Ascaris* with its diet. It is also relatively non-toxic and produced in relatively small amounts by most nematodes studied. It seems probable, therefore, that the compound diffuses out of the animal and does not require a specialised excretory pathway.[139]

5.3.5 Functions of the Excretory System

The excretory system of many nematodes would seem to play a quantitatively minor role in nitrogenous excretion. It may be of greater value in the elimination of other substances but it also has further functions. It should not be inferred that the system is unimportant merely because it has a variable structure and its physiological significance has yet to be demonstrated. However these factors may suggest that it does not have a single universal function in all nematodes. There is sufficient evidence to suggest that the tubular system present in many nematodes differs in function from the associated sub-ventral glands (see figure 1.5).

The excretory system does play a role in the osmotic, and perhaps ionic, regulation of some nematodes. An ampulla in the excretory duct of *Nippostrongylus* and *Ancylostoma* pumps water from the animals in hypotonic solutions.[188] *Ascaris* is known to maintain an ionic content in its pseudocoelomic fluid that differs from that of its surroundings.[171] This ionic regulation may be mediated by the excretory system but does not seem to be associated with a bulk flow of water. Some workers have seen a morphological similarity between nephridia and the nematode excretory system.[189] However the excretory system in nematodes does not seem to utilise the internal body pressure and lacks cilia that are capable of producing a flow of fluid within the system. It has been suggested, from ultra-structural studies, that the lateral canals of the systems in a few nematodes have secretory activity.[139] If this is the case, the excretory system in some nematodes may remove substances from the pseudocoelom and it is worth noting that many bear a relationship to this body cavity similar to the relationship between the malphigian tubules and the haemocoele of insects.

Several animal-parasitic nematodes (*Aspiculuris, Syphacia, Spironoura, Cosmocerca*) with lateral canals in their excretory system, excreted several fluorescent dyes very quickly through the excretory pore after they had been ingested. The dye is not expelled continuously but is emitted as a series of fine jets with pauses between one emission and the next, during which the excretory canals again fill with dye. In animal-parasitic nematodes containing sub-ventral glands as well as the lateral canal system (*Uncinaria* and *Oswaldocruzia*) ingested fluorescent dyes concentrate in the cells of the intestine and do not pass into the pseudocoelom. Fluorescent dyes that have been injected into the pseudocoelom are not excreted and do not become concentrated in the excretory canals. This limited evidence suggests differences in function between dissimilar types of excretory system. Further work may show that one function of the excretory system in some nematodes is to free the pseudocoelom fluid of foreign matter and some metabolic waste.[188,190]

Many nematodes release secretions containing enzymes through the excretory pore and some of these are formed in the sub-ventral glands. This is another example of the structural unity of nematodes: a single opening serves for a range of both excretory and secretory activities. Evidence from a limited number of species suggests that a part of the excretory system is involved in secretory activity accompanying moulting. Unfortunately work of this kind is difficult and the moulting process is poorly understood at present (see section 2.2). Much of the diversity of secretory activity has been reported from animal-parasitic nematodes. Secretions released by the sub-ventral glands from the excretory pore of some species may have an exodigestive function. *Nippostrongylus* releases considerable quantities of an esterase and an aminopeptidase from the sub-ventral and pharyngeal glands and these secretions may be related to communal extracorporeal digestion by aggregations of adults at the mucosa.[11] Trichostrongyle nematodes (for example *Nippostrongylus, Trichostrongylus*) also secrete appreciable quantities of cholinesterase from the excretory pore. The enzyme may inhibit local peristalsis of the host intestine and prevent displacement of the nematode from the mucosa; secondly, it possibly alters the permeability of host cell membranes and results in a leakage of metabolites subsequently assimilated by the nematode.[11,12]

These functions of the excretory system in animal-parasitic nematodes are clearly adaptations to parasitism. One further and distinct example is known for plant parasites. Swollen sedentary species usually secrete a gelatinous matrix around those eggs that are released by the female into the soil and this material is secreted through the excretory pore in *Tylenchulus*.[4]

It would seem that nematode excretory systems have a range of functions. In some an osmotic and ionic regulatory role may be important. The excretory systems in infective juveniles of *Nippostrongylus* and *Ancylostoma* certainly function in osmoregulation and in *Ascaris* the structure possibly aids salt balance.

The role in nitrogenous excretion would seem to be minor in most species but the system may remove various other unwanted material. These broadly excretory functions may frequently be associated with the lateral canal system. Many nematodes also possess sub-ventral glands (see section 1.2.8) which may be mainly responsible for the production of materials secreted from the excretory pore. This orifice probably represents a major outlet for secretions acting on the surface of the animal or on its surroundings. Each species may produce several substances related to its needs in a particular habitat. Over all, the range of secretions released from the excretory pore of nematodes is therefore likely to be substantial.

6 Reproductive Physiology and Hatching

6.1 Reproductive Physiology

6.1.1 The Male System and Spermatogenesis

The reproductive system of most male nematodes consists of a single testis, a seminal vesicle and a vas deferens which opens into a cloaca (figures 1.1 and 6.1), but in some species there are two gonads.[4,10] Most males possess one or

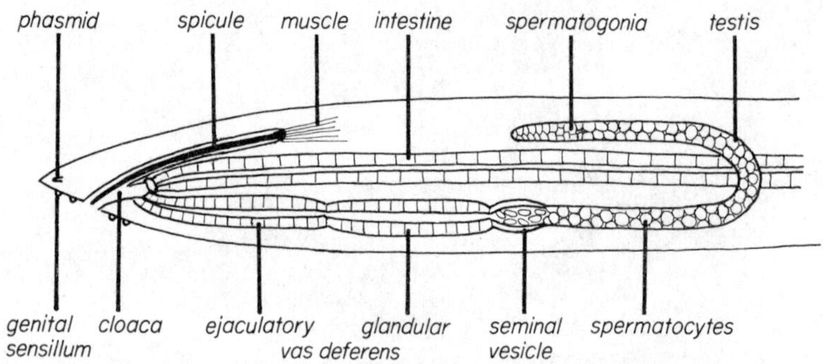

Figure 6.1 The reproductive system of a male nematode.

two copulatory spicules which are formed from hardened cuticle and have neurons running along the centre, terminating in sensory dendrites near the tip of the spicule (see section 9.1.2).[191] The spicules dilate the vulva and vagina of the female during copulation and, in some species (*Heterodera, Dipetalonema*), they also make a channel along which the spermatozoa can pass into the female.[15,191] The sensilla near the tip of the spicule probably allows this hard copulatory aid to feel its way into the reproductive system of the female without damaging her. Males also possess several genital sensilla around the cloacal area (figure 6.2) and in some nematodes the area of the body wall around the cloaca has become expanded to form a copulatory bursa (figure 6.3).[10] The genital sensilla probably help the male to orientate itself prior to, and during, copulation and the bursa is used to grasp the vulval region of the female during copulation.

The testis consists of spermatogonia and spermatocytes within a tube of thin epithelial cells. In telogonic testes the spermatogonia are proliferated from one or more cells at the tip of the testis and this is the commonest system in nematodes. In hologonic testes proliferation of spermatogonia occurs along the whole

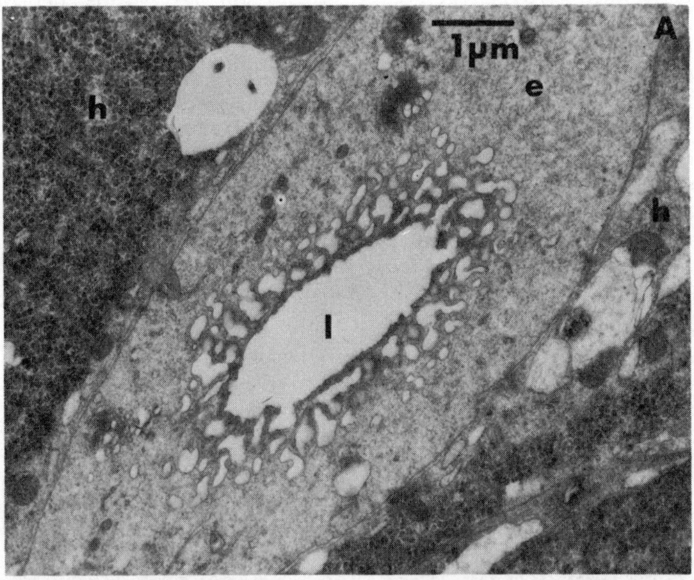

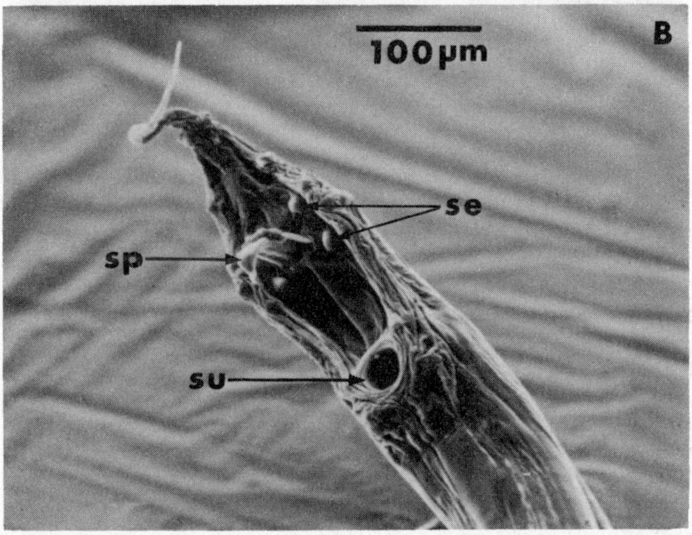

Figure 6.2A Electron micrograph of a section through one of the excretory canals of *Heterakis.* The canals (e) lie in the lateral hypodermal cords of the nematode and are surrounded by hypodermal tissue (h). Note the numerous vacuoles or canaliculi which open into the central lumen (l) of the canal.

Figure 6.2B Tail region of a male *Heterakis* to show sensilla (se) around the cloacal area, the sucker (su) which is used to attach the male to the female during copulation (note the small sensillum within the sucker), and one of the spicules (sp) projecting from the cloaca.

length of the testis (*Trichinella*). Each spermatogonium passes through a phase of growth and differentiation; mitotic divisions form primary spermatocytes and two further maturation divisions give rise to secondary spermatocytes. These undergo reorganisation to form spermatids and ultimately spermatozoa but the nuclear membrane is not reformed after the last division. Four spermatozoa result from the two maturation divisions of the primary spermatocytes.[4,15,92,192]

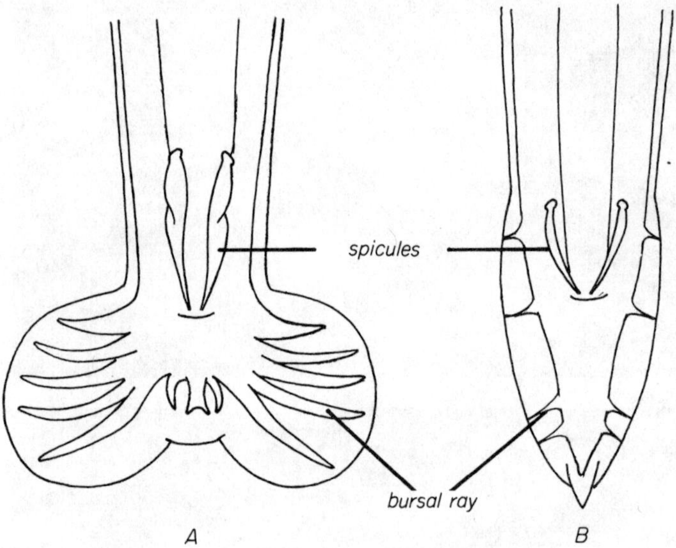

Figure 6.3 The copulatory bursa of *A* a trichostrongyle and *B* a rhabditid (ventral views).

Membraneous organelles containing a fibrous component are present in the cytoplasm of the spermatids of many nematodes; they are formed by Golgi complexes and the granular endoplasmic reticulum of the spermatocyte. The fibrous component of these organelles is released into the cytoplasm when the spermatid develops into a spermatozoon, and is thought to supply the contractile element which brings about pseudopodial movement of the spermatozoon. The nuclear material usually becomes condensed and takes up a definite situation within the spermatozoon (figure 6.4). Microtubules are associated with the spermatozoa of some nematodes and some appear to have a tail, but flagella are not present and all movement is by means of pseudopodia. Mitochondria are present in the cytoplasm of the spermatozoon.[15,192] Spermatozoa are stored in the seminal vesicle of the male until copulation occurs, when they pass along the vas deferens and into the female. The vas deferens is made up of columnar glandular cells, surrounded by a network of muscle in some species.[4,10] Several types of secretion appear to be produced by the cells of the vas deferens but little work has been done on them. 5-hydroxytryptamine occurs in some granules in the vas deferens of *Aspiculuris* and it has been suggested that if this

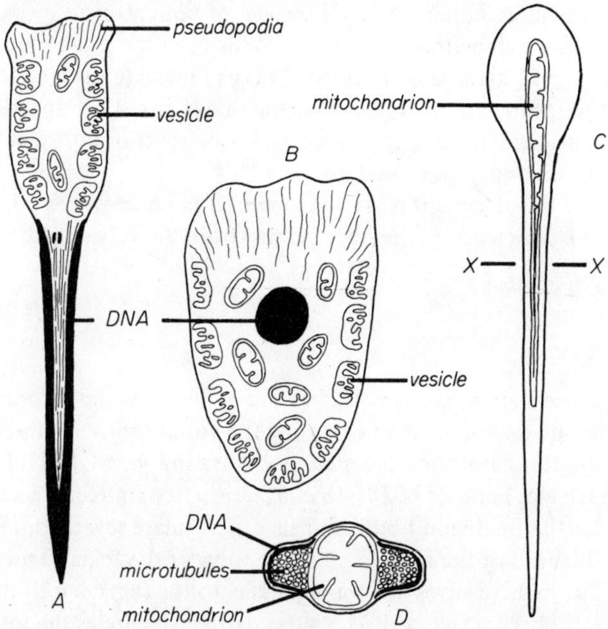

Figure 6.4 Some nematode spermatozoa. *A Nippostrongylus brasiliensis*; *B Heterakis gallinarum*; *C Aspiculuris tetraptera*; *D* transverse section through the 'tail' of *C* at *X–X*. All of these spermatozoa, including those with 'tails' move by putting out pseudopodia.

substance is released with the spermatozoa at copulation it may lead to local contractions of the female system and so aid ascent of the spermatozoa to the seminal receptacle.[15]

6.1.2 Sex Attraction

Adult nematodes produce sex attractants but, in most cases that have been studied, it is the female which produces the attractant and the male which is attracted to the female. In *Panagrolaimus, Trichinella, Camallanus* and *Aspiculuris* both sexes attract each other and presumably both sexes produce sex-specific attractants.[15]

The sedentary females of *Heterodera* attract males which have passed from nearby roots into the soil. The precise source of the attractants in the female is unknown but the chemicals are water soluble and are also slightly volatile. They diffuse through both the aqueous and gaseous phases of the soil and produce a concentration gradient around the female (see section 9.2.1). Four distinct groups of attractants are known, each of which can be characterised by the specificity of certain males to females within the same grouping. Males of some other species respond to more than one attractant and males of

H. rostochiensis will accumulate around females of both *H. avenae* and *H. göttingiana* although neither is attractive to males of the other species. The complex interspecific attraction in *Heterodera* may indicate the course of speciation in this genus but natural cross-attraction is rare. Each species has a distinct host range and males only occur around the roots of host plants which have supported reproduction of that parasite.[16,193]

The source of the attractant is also unknown. It has been suggested that it is derived from the cuticle, from the hypodermis, from the vulva or from caudal glands.[15,16,17]

6.1.3 Copulation

The process of copulation has been studied in relatively few nematodes.[15,4] In most species the male coils the tail around the vulval region of the female and in those species that possess a copulatory bursa this grasps the vulval region of the female. Females of *Ditylenchus destructor* are receptive to males for a week after the final moult but a female can copulate several times during that period. The head of the male touches the vulva of the female and then twists away. The male revolves in a spiral parallel to the female until the spicules and vulva become aligned. The bursa partially encircles the female and holds the two worms together. The spicules are then inserted into the vagina and within a second up to twenty spermatozoa are ejaculated into the posterior portion of the uterus. The male then moves away from the female. During copulation in *Nematospiroides dubius* the worms lie with their heads in opposite directions and the male grasps the female with its bursa. The spicules are inserted through the vulva as far as the ovijector then a colourless substance is released from the vas deferens and forms copulation pads on the outside of the vulva. These may aid in adhesion of the two worms. Spermatozoa pass from the male into the vagina and along the uterus to the seminal receptacle. In *Nippostrongylus* the spermatozoa move along the female system at a rate of about one millimetre in 2.9 hours and after copulation all spermatozoa are within the seminal receptacle after four hours.[15,4]

Changes in size and appearance of the spermatozoa occur within the female, for example the elongate spermatozoa of *Dipetalonema witei* and of *Nematospiroides dubius* become rounded, and a large refringent body appears in the spermatozoon of *Ascaris*. In some species (*Porrocaecum*) spermatids are passed by the male at copulation and only develop into spermatozoa within the female.[15,4,192]

6.1.4 The Female System and Oogenesis

In most nematodes the female reproductive system consists of a pair of genital tubes which open into a common vagina and vulva (figures 1.1 and 6.5).[10] The vagina is lined with cuticle which is continuous with that of the body wall. In many species it is followed by a muscularised ovijector which assists in the expulsion of the eggs from the female. This is followed by the uterus, which may also possess a muscular coat. The cells of the uterus wall contain granular secretory bodies and, in some species, these add an outer coat to the egg as it passes along the uterus (*Ascaris, Thelastoma*). The oviduct extends from the uterus to the ovary and may be modified to form a seminal receptacle. The ovary contains oogonia and oocytes enclosed in a tube of thin epithelial cells.

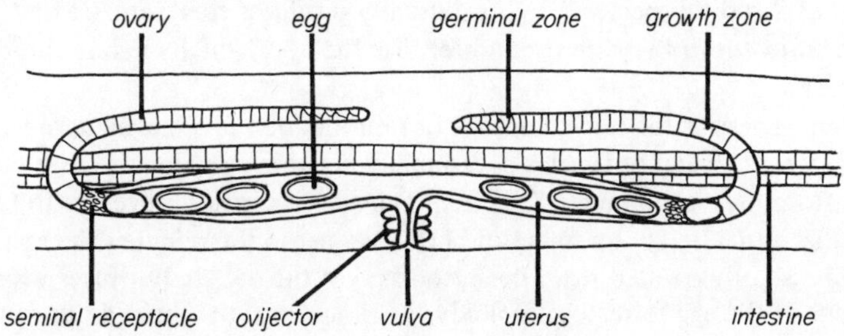

Figure 6.5 The reproductive system of a female nematode.

In telogonic ovaries the oogonia are produced from a germinal cell or cells at the tip of the ovary. The oogonia, and the early oocytes into which they develop, are in cytoplasmic continuity with a central cytoplasmic strand, called the rachis, in many nematodes. The function of this rachis is uncertain but it may perform a nutritive role, it may maintain the basic spatial arrangement of the oocytes, or it may help in synchronising maturation of the germ cells. The oocytes eventually separate from the rachis, grow and become disc-shaped in the ovary. In those nematodes that lack a rachis the oogonia divide to form oocytes which grow in size and pile up as a series of discs in the ovary. Various inclusions appear in the oocyte as it develops; some of these appear to be food reserves for the developing egg but others are egg shell precursors which will be released to form parts of the egg shell after fertilisation.[15,4] Some of these granules contain the unusual esterified glycosides (ascarosides) that are present in a number of nematodes (see sections 4.1.1, 4.2.1 and 6.1.6).

6.1.5 Fertilisation

Syngamy (cross fertilisation) occurs in most nematodes but hermaphroditism occurs in a few species (*Caenorhabditis*) and parthenogenesis in a few others (*Meloidogyne*).[15,19,92]

In syngamy, fertilisation takes place in the seminal receptacle or in the oviduct. The spermatozoa in the seminal receptacle surround the oocyte and one penetrates by removing part of the surface coat of the oocyte so that the membrane of the spermatozoon becomes applied to the oolemma or vitelline membrane. The two membranes fuse and the contents of the spermatozoon enter the cytoplasm of the oocyte. A second membrane then appears beneath the vitelline membrane and the egg shell forms between these two membranes. There is a rapid synthesis of ribosomes around the spermatozoon after entry into the oocyte of *Ascaris* but the pronuclei of the spermatozoon and of the oocyte do not fuse until thirty-eight hours after penetration of the oocyte by the spermatozoon.[15,103]

In hermaphroditic species the gonad usually produces spermatozoa first and these are stored until they are used to fertilise the eggs, which are later produced by the gonad.

Parthenogenesis is the normal means of reproduction in some nematodes (*Meloidogyne javanica*) but males occasionally occur in response to some form of stress from the environment.[18,19] In *M. hapla* two divisions give rise to two polar bodies and a haploid number of chromosomes in the oocyte. The second polar body is not extruded from the cytoplasm of the oocyte but fuses with the pronucleus of the egg to restore diploidy. In *M. javanica* the diploid number of chromosomes is maintained through suppression of chromosome pairing during meiotic prophase and the absence of a second meiotic division. No spermatozoon is needed to activate the oocyte. In females that have been inseminated by the occasional male the spermatozoon enters the ovum but then degenerates.[15,19]

In most nematodes development of the egg occurs only after it has been penetrated by a spermatozoon. In a few species this serves only to activate the egg, as no fusion of nuclear material occurs and the pronucleus of the spermatozoon degenerates in the cytoplasm of the egg. This pseudogamy occurs in *Rhabditis maupasi* and the somatic number of chromosomes is restored by a mitotic division of the oocyte pronuclei.[15,92]

6.1.6 Structure and Formation of the Egg Shell

The eggs of most nematodes are ovoid and similar in size, but variations in form and structure do occur (figure 6.6).[10] The egg shell usually has three main layers: an inner layer, which is sometimes called the lipid or ascaroside layer; a thicker, chitinous layer that forms the main part of the shell; and a thin, outer layer that originates from the vitelline membrane or the oolemma of the fertilised oocyte (figure 6.7).[4,15,103] All of these layers are formed by the egg. An additional outer covering may be secreted by the wall of the uterus. This layer may give a sculptured surface to the egg and may even be drawn out into processes (figure 6.6). In some species the uterus secretes a glutinous material and the eggs adhere to a substrate when laid.

After fertilisation of the egg a new membrane forms beneath the original vitelline membrane and the egg shell forms between these two membranes

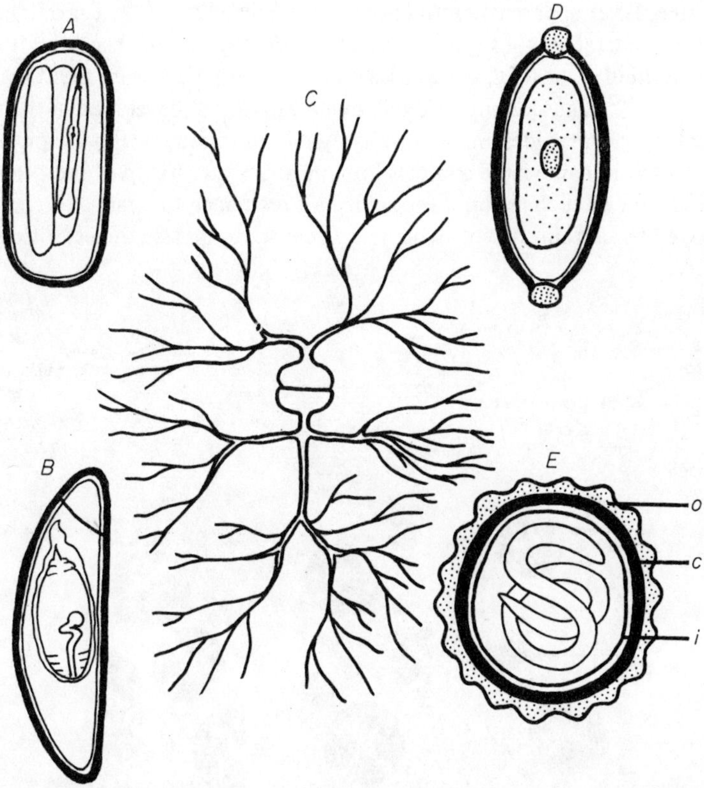

Figure 6.6 Eggs of some nematodes. *A* tylenchid egg containing a juvenile; *B* oxyuroid egg (*Leidynema*) containing a contracted infective juvenile; *C* egg of *Mermis*; *D* unembryonated egg of *Trichuris*; *E* Egg of *Ascaris* containing an infective juvenile. *c* chitin–protein layer; *i* inner (ascaroside) layer; *o* outer (uterine) layer.

(figure 6.7*A*).[4,15,103] The main part of the egg shell is the chitinous layer. This is the only structure in nematodes known to contain chitin and it gives rigidity to the egg shell. In some species the absence, or reduction in thickness, of this layer at one or both poles is associated with opercula in these regions (figure 6.6). The layer consists mostly of chitin with a little protein in the Ascaroidea and the Oxyuroidea, but the egg shell of *Heterodera rostochiensis* contains much less chitin than protein.[15,135] Some of this protein is in the form of collagen. Material for this layer is formed in the granular endoplasmic reticulum around the periphery of the newly fertilised egg and is secreted into the space between the original vitelline membrane and the new membrane of the egg (figure 6.7*B*). The inner layer of the shell is then formed by a movement of refringent granules towards the periphery of the cytoplasm, where they are extruded into the space between the chitinous layer and the membrane of the egg (figure 6.7*C*).[4,15]

The inner layer apparently influences the permeability of the egg and has the permeability characteristics typical of a lipid layer but, in some species, it consists of esterified glycosides, called ascarosides because they were first identified in *Ascaris.*[101,102] Unusual impermeable qualities are well developed in the eggs of a number of nematodes and are probably due to this layer. For example, the eggs of *Ascaris* remain viable and develop in many tissue fixatives and other noxious chemicals such as sulphuric acid. This resistance to a range of chemicals is destroyed by certain solvents which apparently disorganise the esterified glycosides.

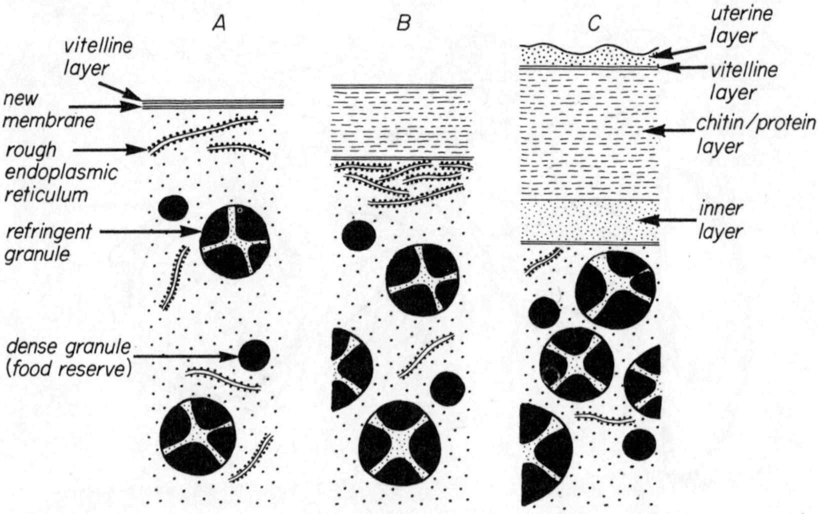

Figure 6.7 Stages in the formation of the egg shell of *Heterakis gallinarum. A* immediately after fertilisation (note the formation of a second membrane beneath the original vitelline membrane); *B* formation of the chitin–protein layer between the outer and inner membranes; it is produced by material that originates in the rough endoplasmic reticulum that congregates around the periphery of the egg; *C* formation of the inner (ascaroside) layer and deposition of the outer, uterine, layer. Refringent granules (containing the ascarosides) move through the cytoplasm and discharge their contents across the egg membrane to form the inner layer of the egg shell. The uterine layer is deposited on the outer surface of the egg as it passes along the uterus.

Some nematodes are ovoviviparous and the egg shells of these species often have a reduced complexity (*Panagrellus, Turbatrix, Trichinella* and various filarial nematodes). In these nematodes the larvae hatch within the uterus and subsequently emerge from the female as active juveniles. Presumably the female offers a degree of protection to the egg during embryogenesis and provides a suitable medium for rapid development.

The eggs of *Heterodera* are protected within a cyst, which is formed by the tanning of the cuticle of the dead female. This tanning process involves a polyphenol oxidase, which probably produces a quinone capable of hardening the cuticular protein.[31] Some plant-parasitic nematodes surround their eggs with a protective gelatinous mass (*Meloidogyne, Tylenchulus, Rotylenchus,* and several species of *Heterodera*). This mass is secreted by the rectal glands in *Meloidogyne,* by the excretory system in *Tylenchulus,* and is probably produced by the uterus in *Heterodera.*[4] These gelatinous masses and the cyst wall of *Heterodera* species may reduce predation by soil organisms and they may also retard the loss of water from eggs in a dry soil.

6.1.7 Egg Laying

Movement of the egg within the reproductive tract is brought about by a network of muscle around the uterus and/or as a result of the kneading action of body-wall muscles. Many nematodes possess a network of muscles around the uterus and also have a muscular ovijector.[10] In these species the eggs are moved along the uterus by the peristaltic-like action of the uterus until they enter the ovijector. Here more powerful, circularly arranged muscles squeeze the egg through the ovijector and expel it through the vagina.
ovijector and expel it through the vagina.

Other nematodes lack a muscular uterus and egg transport is probably similar to that described for *Aphelenchoides blastophthorus.*[194] In this species the eggs are moved down the ovary by the pressure of younger eggs increasing in number and size behind, and by the kneading action of the body wall muscles. Once the egg shell has formed, alternate shortening of the front and hind regions of the nematode result in movement of the egg along the uterus. Contraction of somatic muscles increases body pressure, compresses the uterus and forces the egg out through the vagina and vulva when the dilator muscles contract. The egg is deformed during its passage through the vulva.

6.2 Hatching

6.2.1 Emergence of Juveniles from Eggs

Many animal-parasitic nematodes hatch outside their hosts and have free-living stages that subsequently develop into infective juveniles (*Ancylostoma, Nippostrongylus, Trichostrongylus*). During hatching of *Trichostrongylus retortaeformis* the innermost layer of the egg shell is first disrupted by enzymes and by the free movement of the juvenile within the egg. The egg shell becomes permeable and the subsequent uptake of water by the juvenile increases the turgor pressure within the animal, causing it to become relatively immobile. During this period the volume of the nematode increases until sufficient pressure is exerted on the weakened egg shell to rupture it.[195]

Some animal-parasitic nematodes (*Ascaris, Parascaris*) do not have free-living juveniles and hatching is suppressed until the egg is ingested by a suitable host.

Hatching of *Ascaris* eggs is initiated by a combination of parameters that are characteristically found in the gut of its host (section 6.2.4). After stimulation the juvenile releases a hatching fluid containing a number of enzymes; these include an esterase capable of disrupting the inner ascaroside layer, a chitinase and a protease, which dissolve a portion of the chitinous layer of the egg shell. The juvenile then protrudes from this hole, still surrounded by a membrane, which is eventually ruptured or digested.[34]

The hatching of *Meloidogyne javanica* also shows some similarities to that described for *Trichostrongylus retortaeformis.* The juvenile secretes enzymes that destroy the inner lipid layer of the egg and it also exerts pressure on the poles by means of locomotory activities.[4] Juvenile tylenchids possess a stylet, and some species of *Meloidogyne* and *Heterodera* use this to cut a slit in the egg. *Heterodera rostochiensis* differs from *Meloidogyne javanica* in requiring a chemical stimulus from host roots for an appreciable rate of hatch to occur (section 6.2.4). Activity of the juvenile of *Heterodera rostochiensis* in the egg gradually increases following stimulation, and secretory activity occurs within the dorsal pharyngeal glands. After a period of movement within the egg the juvenile carries out a remarkable behavioural sequence. The head is braced against one pole of the egg and the stylet thrusts through the egg shell (figure 6.8). The juvenile repeats this process many times, producing a continuous cut. Precise sensory and motor co-ordination is involved and the animal will locate and recut regions of the slit that have an incomplete perforation. The animal emerges from the egg once a large enough slit has been produced in the egg shell.[196] Juveniles of *Heterodera rostochiensis* almost fill the space within the

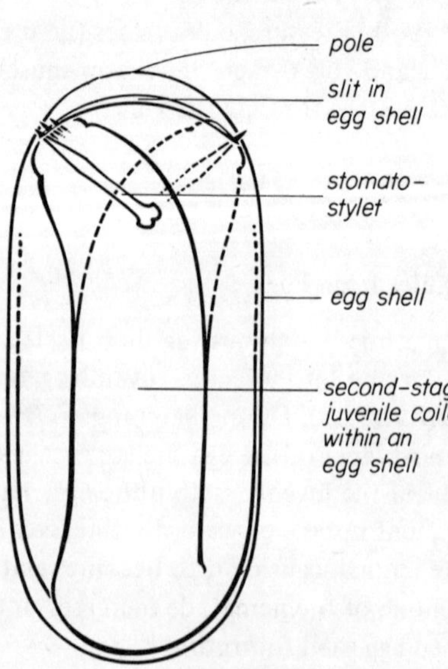

Figure 6.8 A juvenile of *Heterodera rostochiensis* cutting a slit in a pole of its egg shell prior to hatching. (After Doncaster and Shepherd.[196])

egg and in this species the uptake of water by the juvenile occurs mainly after hatching.[197]

6.2.2 General Environmental Effects on Hatching

The development of the juvenile within the egg and its subsequent emergence from it are influenced by external factors such as the availability of moisture, the oxygen content of the surroundings and the ambient temperature. Some oxygen is essential for embryogenesis in most animals and nematodes are unlikely to be an exception. Hatching also requires some oxygen in most species; hatching of *Meloidogyne javanica* is reduced in conditions where the amount of oxygen present is low, and this may be typical of many nematodes from aerobic habitats[151] (figure 6.9). Some animal-parasitic nematodes do not have a free-living juvenile stage but hatch within the gut after ingestion by a host. Eggs of these species (*Ascaris, Parascaris*) are unusual because they are adapted to hatch in the very low partial pressures of oxygen that are characteristic of the gut lumen of their hosts.

A low oxygen concentration in soils is particularly associated with wet conditions because water replaces air in the soil pores and reduces the rate of

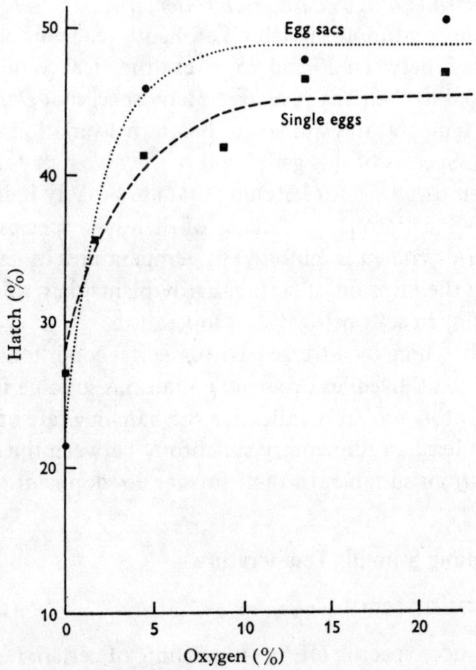

Figure 6.9 The percentage hatch of juveniles of *Meloidogyne javanica* from single eggs and from egg sacs at different oxygen concentrations. (After Baxter and Blake.[151])

diffusion of oxygen through the soil. This reduction in oxygen supply decreases the availability of oxygen to the nematode but the precise relationship will depend upon many factors including the soil structure, and its microflora and microfauna. The over-all effect of changes in soil moisture on hatching can be seen for juveniles of *Heterodera schachtii* emerging from cysts in sandy soils. Very wet soils suppress hatching because they have a reduced oxygen content. The rate of hatching increases in less moist soils but is inhibited by very dry conditions,[198] which probably reduce the water content of the eggs.

Lack of water in soil also affects hatching of *Ostertagia.* Eggs released in the faeces of the sheep in early summer subsequently hatch and the juveniles produced infect lambs later in the same season. Development of the juvenile and hatching of the egg require the presence of some oxygen and a suitable ambient temperature, but they are also dependent on the accumulative effect of exposure to moist conditions for about twenty-five days. In a typical summer such damp conditions are not continuously available in the surface layers of soils occupied by the eggs and, therefore, the maximum rate of hatching does not occur until eight to ten weeks after egg deposition has occurred. Hatching of the infective-stage juveniles of *Nematodirus battus* in spring is reduced by lack of oxygen in wet soils and also by dry conditions.[199] Soil moisture is probably important because uptake of water is frequently associated with the hatching process in nematodes (*Trichostrongylus, Ascaris, Heterodera*).

The optimum temperature for hatching of many free-living and plant-parasitic nematodes is between 20 and 25 °C but the ideal conditions for some species can be related to their environment. Many species of *Heterodera* hatch best in fluctuating temperatures similar to the diurnal variation that occurs in the surface of soils. Species of this genus and of *Meloidogyne* tend to show a range of optimal temperatures for hatching that are broadly indicative of the geographical region and seasonal preference of their host species.[200] Hatching of most species of *Heterodera* is inhibited by temperatures of less than 8–10 °C and this can reduce the invasion of certain hardy plants that are capable of growth in early spring in soils below this temperature.

Suppression of hatching by adverse environmental conditions ensures that maximum emergence of juveniles occurs in conditions suitable for an active nematode. These factors probably influence the hatching rate of most nematodes and enable at least a rudimentary synchrony between the emergence of juveniles and conditions suitable for their further development.

6.2.3 Specific Hatching Stimuli: Temperature

6.2.3(a) Animal-parasitic nematodes

Temperature has a more specific effect on hatching of certain nematodes than that just described. Eggs of *Ascaris* and *Aspiculuris* hatch in response to temperature and a number of other factors after they have been ingested by a host.

Juveniles will not hatch *in vitro* below 30 °C but emerge more readily as the temperature is raised towards that of their homeothermic hosts.[34] A response to temperature of a different nature is involved in the hatching process of *Nematodirus battus.* Adults of this species do not usually survive in lambs from one season to another. The persistence of a population in a locality is dependent on the eggs, which pass out in the faeces of the host during the summer. Development in the egg proceeds to the third-stage juvenile on the pasture during the summer and autumn, but they remain dormant within the egg throughout the winter. Hatching is initiated in the spring by a rise in the soil temperature from low values during winter, and this is coincident with the presence of young lambs on the pasture. The temperature response is quite precise for the population as a whole and an empirical formula based on soil temperature and moisture from mid-summer to the following spring has proved of practical value in anticipating the timing and the size of the annual hatch of juveniles in northern Britain.[199] This response to temperature is similar to the diapause of some insects and has an obvious survival value for this nematode.

6.2.3(b) Plant-parasitic nematodes

The cereal cyst nematode *Heterodera avenae* also responds to changes in soil temperature in a manner that is similar to diapause in some insects. In Britain, this nematode hatches in temperatures characteristic of spring if first exposed for several weeks to temperatures similar to those in soils during winter. Populations from Australia show a somewhat different response. The dry soils during summer on that continent inhibit hatching of *H. avenae,* but a fall in temperature during autumn favours emergence of juveniles. Winter cereals are invaded and the cysts develop by early summer.[200] The difference in response to climate in these two populations has adaptative significance, for it ensures that the hatching of juveniles is synchronised with the availability of host plants.

6.2.4 Specific Hatching Stimuli: Chemical

6.2.4(a) Plant-parasitic nematodes

Most species hatch without delay once the juvenile has reached a certain stage of development in suitable environmental conditions. Many species of *Heterodera* are exceptions to this and in these nematodes hatching is markedly stimulated by substances released into the soil by certain plants. Root diffusates collected from soils containing any of a range of hosts enhance the number of juveniles that hatch from within the cysts of *Heterodera schachtii* under laboratory conditions. Some juveniles of this species hatch in water and they also respond to diffusates of certain non-host plants. *Heterodera rostochiensis* has more specific hatching requirements, for few juveniles hatch in water and an appreciable number only respond to diffusates from potato, tomato and a few other solanaceous host plants. *Heterodera göttingiana,* the pea cyst nematode, shows an even more distinct response because juveniles will not hatch in laboratory experiments. They emerge only in field conditions and are stimulated to hatch by the growing of peas in the soil.[201] The number of viable eggs declines on an annual basis for all field populations of *Heterodera* but species with narrower and more precise host ranges and more precise hatching requirements do tend to be more persistent.[202] This would seem to have an adaptative significance, for those species with a wide host range of common plants are able to multiply more frequently than are those that can only reproduce on a limited range of host plants.

The emergence of juveniles of *Heterodera* is a complex event; the cyst may act as an ecological unit and the hatching of eggs it contains may be limited by an interaction of factors within the cyst. Hatching of *H. rostochiensis* has been more fully studied than that of other species. In field conditions a proportion of the eggs of this nematode hatch each spring and the juveniles subsequently die in the absence of host plants. Substances released by soil organisms are partly responsible for this and viable eggs in cysts of *H. rostochiensis* show an approximately exponential decay at a rate usually between 30 and 50 per cent per annum.[202] The stimulus released by host roots has two main functions: it releases more juveniles than would hatch in the absence of the host and the hatch from each cyst occurs over a relatively short period, when roots suitable for invasion are near the cyst.

The hatching factor for *H. rostochiensis,* which is released by tomato and potato roots, has been extensively studied because of its possible application in the control of this nematode. Unfortunately, the active chemical is relatively unstable in soil conditions. It has proved difficult to purify and lacks a readily identifiable structure. Further work is required but at present it is widely believed to have an empirical formula of approximately $C_{18}H_{24}O_8$. It is apparently acidic in nature, it may have a lactone group and perhaps two hydroxyl groups that are spaced at a known distance from each other. It may have a biological activity similar to certain cardiac glycosides and is fully active as a hatching agent at a concentration of five to ten parts in 10^8. A number of artificial hatching agents cause *H. schachtii* to emerge from the egg, but picrolonic acid, anhydrotetronic acid and sodium metavanadate are the main effective artificial hatching agents for *H. rostochiensis.* Sodium metavanadate has a somewhat different action from the other artificial hatching agents as it softens the egg shell. No hatching agent has yet proved economically effective in complete or integrated control of this animal.[201]

The mode of action of the potato root hatching factor is not known. Its addition to cysts of *Heterodera rostochiensis* does increase their oxygen consumption within twenty-four hours and this is accompanied by a change in the adenylate energy charge of the cyst contents[203,204] (figure 6.10). Both these changes suggest that the metabolism of unhatched juveniles is altered shortly after stimulation. The action of the diffusate may involve a change in the permeability of the egg shell initiating a sequence of events that finally results in emergence of the juvenile (section 6.2.1). A similar sequence could also occur if the egg shell is freely permeable to the hatching factor and the nematode responds to its presence within the egg. A less likely alternative is that the chemical released by host roots is a metabolite required by the juvenile,

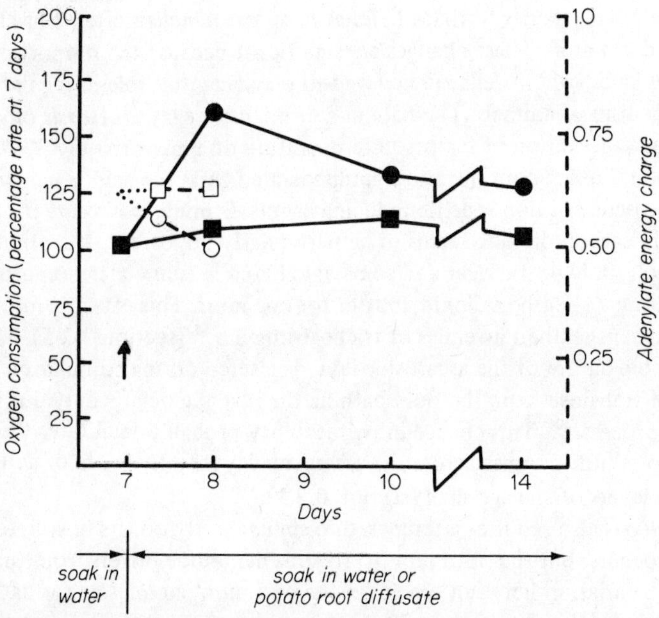

Figure 6.10 Changes in oxygen consumption and energy charge of eggs of *Heterodera rostochiensis* after addition of potato root diffusate to individual cysts.

- ● oxygen consumption of cysts soaked in potato root diffusate.
- ■ oxygen consumption of cysts soaked in water.
- ○ adenylate energy charge of cysts soaked in diffusate.
- □ adenylate energy charge of cysts soaked in water.

$$\text{Energy charge} = \frac{1}{2}\left\{\frac{2\,[\text{ATP}] + [\text{ADP}]}{[\text{ATP}] + [\text{ADP}] + [\text{AMP}]}\right\}$$

(After Atkinson and Ballantyne.[203,204])

and which is not present in the egg.[34] Understanding the mechanisms involved in hatching of species of *Heterodera* is of great economic significance. However the isolation of a hatching factor for any economically important species of *Heterodera* may prove of even greater value if this were to result in the synthesis of a cheap analogue chemical which was useful in controlling these nematodes.

6.2.4(b) Animal-parasitic nematodes

In some animal-parasitic species the egg must be ingested by the host and the juvenile hatches within the alimentary system (*Ascaris, Toxocara, Leidynema, Mermis*). Eggs of these nematodes have to develop to an infective stage before they will respond to ingestion. During embryonation the egg of *Ascaris* rapidly consumes oxygen and lipid but this rate of utilisation falls as the infective egg enters dormancy.[109] The adenylate energy charge and ATP/ADP ratio of *Ascaris* eggs increases to a high level in the dormant period[205] (figure 4.6) but falls in a similar manner to that of *Heterodera rostochiensis* after it has been stimulated to hatch. These changes are significant because the proportions of ATP, ADP and AMP in cells are believed to play a central role in regulating the metabolism of animals. The hatching of infective eggs of *Ascaris in vitro* requires a combination of factors: a temperature of approximately 37 °C, a high carbon dioxide concentration, undissociated carbonic acid, a neutral pH and the presence of non-specific reducing agents. Stimuli *in vivo* are probably similar and carbon dioxide seems to be particularly important. Its action is that of a physiological trigger and some eggs hatch in saline after exposure to a bicarbonate–carbon dioxide buffer for one hour. This effect is more marked for ensheathed juveniles of trichostrongyles[34] (section 2.2.5). The impermeable nature of the ascaroside layer is altered during stimulation because leakage of trehalose from the fluid bathing the juvenile occurs only during the hatching process.[206] This change in permeability probably facilitates the passage of hatching fluid enzymes across the ascaroside layer to their site of action in the outer layers of the egg shell (section 6.2.1).

Aspiculuris also requires a temperature similar to that of its host before hatching occurs, but the optimum pH for this nematode differs from that for *Ascaris.*[34] Variation between species in the optimum parameters for hatching increases the probability that each responds only to ingestion by a suitable host and ensures that the juveniles are released in a preferred region of the alimentary tract.

The specific hatching mechanisms of animal- and plant-parasitic nematodes are a major adaptation to parasitism in these animals. Dormant eggs are capable of surviving for prolonged periods in conditions that would be unfavourable to the hatched juvenile. The relatively ephemeral infective juveniles are mainly released in conditions that are suitable for invasion of the host required for further development of the nematode. Hatching specificity also increases the degree of synchrony between reproduction of the parasite and the availability of further susceptible hosts; the ensheathed juveniles of some nematodes have a similar adaptative significance (section 2.2.5).

7 Neuro-muscular Physiology

7.1 Muscles

The somatic musculature of nematodes consists of a single layer of longitudinally orientated, spindle-shaped cells that interdigitate with each other. These muscle cells are grouped into rows (quadrants) between the dorsal, ventral and lateral cords (figure 1.2). Each muscle cell has a contractile region, which contains the myofibrils, and a non-contractile sarcoplasmic part which contains the nucleus, mitochondria, glycogen and lipid (figure 1.4). The ribbon-like contractile part of each cell is shaped like a trough, which may be wide and shallow (platymyarian), narrow and deep (coelomyarian) or may completely encircle the sarcoplasm (circomyarian) (figure 7.1).[4,207] Circomyarian muscle is usually associated with

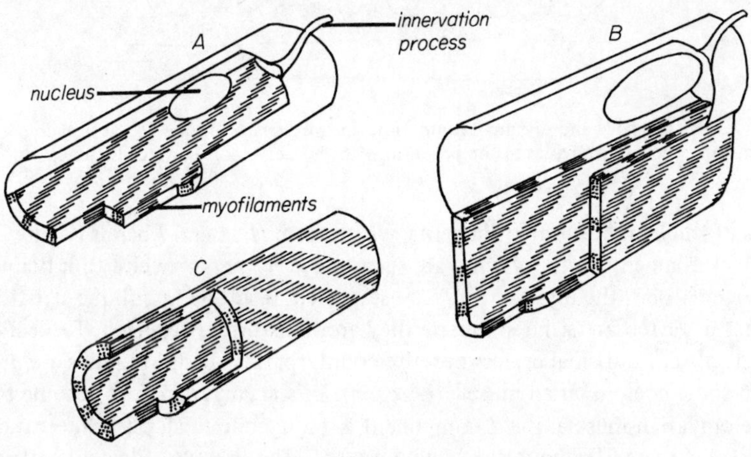

Figure 7.1 Cut-away diagrams showing different types of obliquely striated muscle cells in nematodes. *A* Platymyarian; *B* coelomyarian; *C* circomyarian. (Redrawn from Hope.[207])

specialised muscles. Unlike the muscles of most animals, those of nematodes are not innervated by motor nerves, instead a slender innervation process extends from the non-contractile part of the muscle to one of the longitudinal nerves of the body. Here the innervation process divides into several finger-like processes, which make synaptic contact with the nerve fibres (figure 7.2).[158,208]

The somatic muscles of those nematodes that have been examined in detail (for example, *Ascaris, Deontostoma, Caenorhabditis*) are similar to certain muscles of annelids and the adducta muscle of molluscs in that they are obliquely striated.[207,208] The contractile portion of the muscle cell contains interdigitating

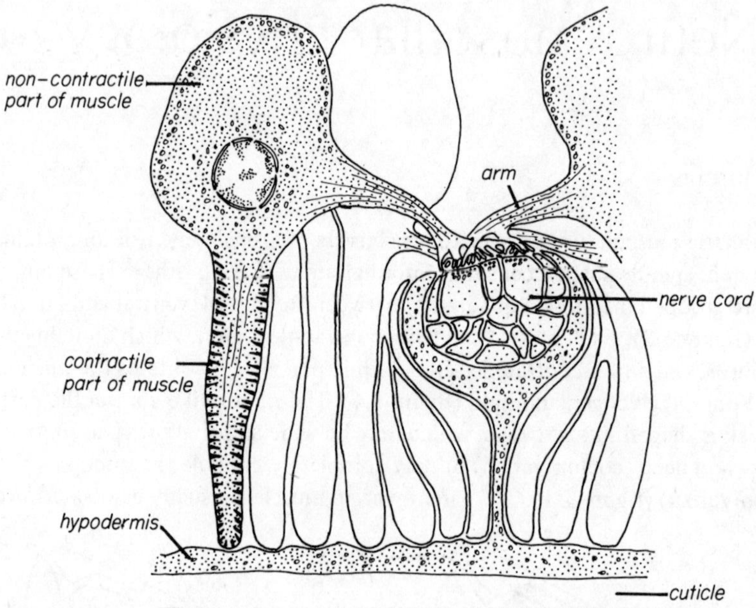

Figure 7.2 Muscle cells and myoneural junctions in transverse section. The arm of the innervation process subdivides at the junction with the nerve cord. (After Rosenbluth.[208])

arrays of thick and thin myofilaments, which form *H, A* and *I* bands (figure 7.3). The thick filaments in the *A* band are surrounded by ten to twelve thin filaments but in nematodes the myofilaments are staggered, resulting in oblique striations, whereas in vertebrate striated muscle they are transversely arranged. Typical *Z* lines are absent but smaller, less orderly, counterparts called *Z* bundles are present in some coelomyarian muscles (*Ascaris*). In platymyarian, and in some types of coelomyarian muscle, the Z component is a long, elliptical, plate-like structure called the *Z* plate (*Trichodorus, Longidorus*).[209] The thin myofilaments attach to these *Z* bundles or *Z* plates. Plasma membrane is associated with the *Z* bundles in *Ascaris* and, together with flattened, membranous cisternae, form dyads and triads. It is thought that these may be strategically located, low impedance areas through which local currents are selectively channelled.[7]

It is probable that the thick filaments present in nematode muscle cells contain myosin and that the thin filaments contain actin as both substances are present in these muscle cells. Paramyosin is also present and is thought to form the central core of the myosin helix.[56]

During contraction of vertebrate striated muscle the actin filaments slide past the filaments of myosin, resulting in an increase in their overlap and a decrease in width of the *I* band. Perfusion of *Ascaris* muscle preparations with piperazine citrate and with acetylcholine has shown that during contraction the filaments slide past each other, as in vertebrate striated muscle, but there is also an increase in the degree of overlap of the obliquely arranged, thick filaments and a decrease in the extent of their stagger, so that the *Z* bundles become more nearly transverse. This results in folding and coiling of the intracellular skeletal fibres of the muscle cell.[210]

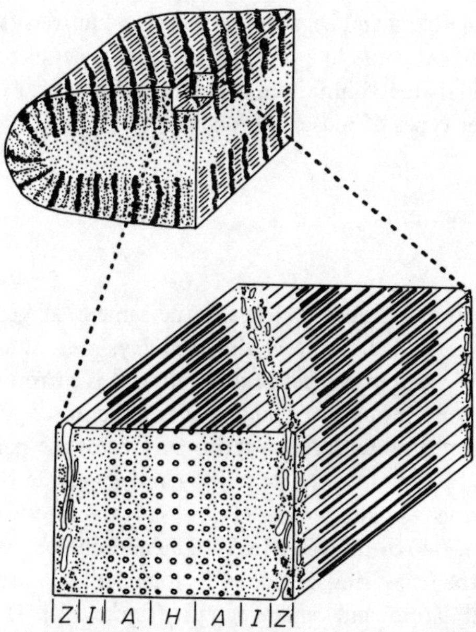

Figure 7.3 Part of a muscle fibre showing the pattern of striations in three planes. Note the obliquely striated arrangement of the myofilaments and the positions of the *H, A, I* and *Z* bands. (After Rosenbluth.[7])

Changes in length of up to 15 per cent occur during feeding and defecation of some nematodes and this involves changes in length of the muscle fibres. This should alter the tensions that they develop, but the muscle tensions are able to be maintained because of the obliquely striated nature of the muscle cell.[9]

While these somatic muscle cells structurally resemble vertebrate cross-striated muscles – except that the cross striations are oblique instead of transverse – they differ from vertebrate striated muscle in that they are spindle-shaped fibres, contraction and relaxation is slow, the cells are spontaneously active and the myosin ATPase is similar to that of certain smooth muscles.[9] They are similar to many invertebrate muscles in that they contain paramyosin.[56]

The pharyngeal musculature is radially arranged and is a single sarcomere in length in those nematodes that have been examined (*Enoplus, Caenorhabditis*). The thin myofilaments are arranged around the thick myofilaments as in the somatic muscles but there are only six thin filaments about each thick filament. The thick filaments have a prominent core, which probably contains paramyosin.

The bundles of myofilaments are interspersed with mitochondria and a network of vesicles, which are probably part of the sarcoplasmic reticulum.[55,56]

Specialised muscles are associated with various regions of the alimentary tract and the reproductive system but little is known about the detailed structure and physiology of these muscles. They are used to dilate the stoma, the vagina and the anus, to move the mouth stylet, the copulatory spicules and copulatory bursa, to move spermatozoa and eggs along the reproductive system, to move gut contents along the intestine in a few species, and to act as sphincter muscles such as at the intestinal–rectal junction. Many of them appear to be circomyarian in structure but other types of muscle do occur.[10]

7.2 The Nervous System

7.2.1 Structure

The nervous system of nematodes is basically the same in all genera. A circumpharyngeal commissure called the nerve ring is always present and consists of nerve fibres with a few supporting cells (glial cells). It is surrounded by connective tissue, which is continuous with the hypodermal cords. Six nerves, which arise from ganglia associated with the anterior surface of the nerve ring, divide into branches in the lip region of the nematode and innervate the inner labial, the outer labial and the cephalic sensilla (figure 7.4) (section 9.1.2).[4,10,59] A small dorsal ganglion, two or more lateral ganglia and one or two ventral ganglia are associated with the nerve ring. Six or eight longitudinal nerves run posteriorly from the dorsal, lateral and ventral ganglia (figure 1.2). They are closely associated with the hypodermis and are often almost completely surrounded by hypodermal tissue (figure 7.2).[208] The ventral nerve is embedded in, or runs alongside, the ventral cord and is the largest nerve. It is a paired structure for some of its length and it is also ganglionated at intervals. A rectal ganglion is associated with a pair of rectal commissures, which arise from the ventral cord, before it terminates in the pre-anal ganglion, and which encircle the rectum or cloaca. This pre-anal ganglion is connected to the lumbar ganglion by means of commissures (figure 7.4). The dorsal nerve arises from the dorsal ganglion and runs alongside, or is embedded in, the dorsal cord. It is smaller than the ventral nerve and is not ganglionated along its length. At the posterior end of the nematode it bifurcates and joins the lumbar ganglia.[4,10,189] The lateral ganglia are the largest ganglia associated with the nerve ring. A nerve runs forward from each lateral ganglion to innervate the amphids (section 9.1.3) at the anterior end of the nematode. Two main, lateral nerves extend from the lateral ganglia to the lumbar ganglia at the posterior end of the nematode. A few ganglia occur along the length of these nerves. Many nematodes also have sub-dorsal and sub-ventral nerves extending posteriorly from the lateral ganglia. All of the nerves that run posteriorly from the nerve ring are connected to each other by commissures at intervals along their length.[4]

The nerve ring, with its associated nerves and ganglia, is thought to be the co-ordinating centre of the nervous system. It also controls the main locomotory activities of the nematode. The dorsal nerve is thought to be chiefly motor, the lateral nerves mainly sensory, and the ventral nerve partly motor and partly sensory in function but this is based more on their anatomy than on physiological evidence. The collection of nerves and ganglia at the posterior end of the nematode (figure 7.4) is apparently concerned with the control of defecation, with copulation in the male (innervation of sense organs, spicules and the musculature) and with reverse locomotion.[4]

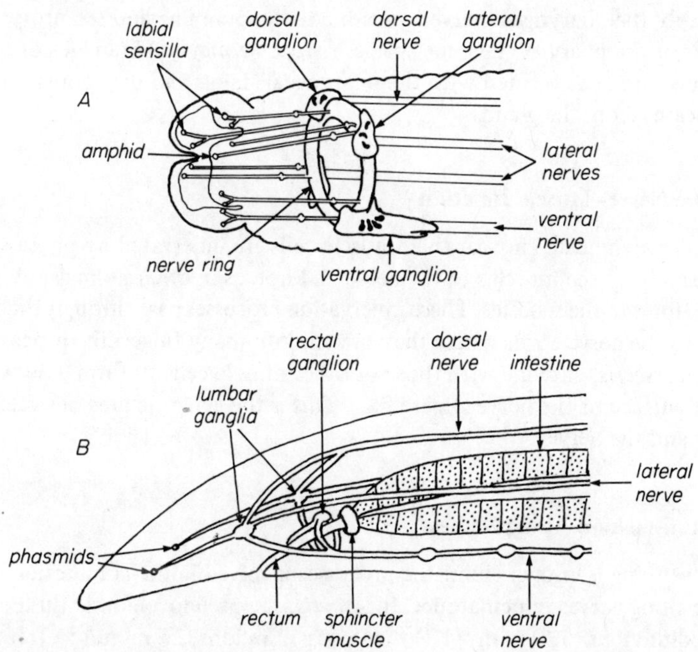

Figure 7.4 The nervous system of a female *Ascaris. A* Head region; *B* tail region. (*A* after Davey, *Amer. Zool.* 6 (1966); *B* after Crofton.[189])

Several large, ventro–lateral commissures occur in the hypodermis of some free-living and plant-parasitic nematodes and are known by the following names. Cephalids occur in the cephalic region and the cuticle overlying them is greatly reduced in thickness. The hemizonid is a commissure that joins the sub-ventral nerve with the lateral ganglion and is situated near the excretory pore. The hemizonion is a small commissure that lies behind the hemizonid. The caudalids are anal–lumbar commissures that link the pre-anal ganglion to the lumbar ganglion. Some of these commissures may respond to pressure on the nematode. The hemizonid may act as a sensory receptor associated with the excretory duct and may also be associated with neurosecretory activity during moulting.[4]

There is a nerve ring in the posterior end of the pharynx and three nerves extend from it to innervate each sector of the pharynx. The nerves are embedded in the pharynx but are in continuity with the pseudocoelom by means of mesaxons.[58] Connections between this nervous system and the nerve ring have been detected in *Ascaris lumbricoides* and in *Caenorhabditis elegans*. Two neurons pass from the lateral cord into the anterior end of the pharynx where they end in an electron-dense specialisation adjacent to two pharyngeal neurons. The two neurons which enter the pharynx are interneurons that receive synaptic input in the main nerve ring of the nematode and are afferent to the pharynx. They are the only two neurons that connect the somatic nervous system to the pharynx.[59] Presumably the pharyngeal nerves co-ordinate the pumping and secretory activities of the pharynx. In some species (*Nippostrongylus*) branches of the nerves are closely associated with the pharyngeal glands and may control release of secretions from the glands.[211]

7.2.2 The Nerve–Muscle Junction

Nematodes are unusual in that their muscle cells are innervated by processes which pass from the muscles to the nerve and not, as in other animals, by nerves sending fibres to the muscles. These innervation processes pass through the pseudocoelom to the nerve cords where they branch into many finger-like processes, which, in *Ascaris*, entwine with those of other muscle cells to form a network over the surface of the nerve (figure 7.2). This is the synaptic area between the muscles and the nerves.[158,208]

7.2.3 Transmission Along Nerves

Almost nothing is known about the processes involved in the conduction of an impulse along nerves in nematodes. In *Ascaris*, the pseudocoelomic fluid contains more sodium ions (129 m mol l^{-1}) than potassium ions (25 m mol l^{-1}). It is probable, therefore, that the action current arises from an influx of sodium ions and the resting potential depends on an excess of potassium ions in the cytoplasm, as occurs in other animals.

7.2.4 Neuro-muscular and Synaptic Transmission

Acetylcholine is an important chemical transmitter substance in the nervous system of most animals, including nematodes. Nerves that secrete it at synapses and muscle end-plates are referred to as cholinergic nerves. Acetylcholine or acetylcholine-like substances have been isolated from a number of species of nematode and appears to be concentrated in the nervous system, as the region of *Ascaris* that contains the nerve ring contains fifteen times more acetylcholine per unit weight than strips of body wall.[158]

Acetylcholine causes contraction of the muscle of *Ascaris* when applied to isolated strips of body wall and relaxation occurs when the drug is washed away. The extent of the contraction is directly related to the concentration of acetylcholine used. Eserine, which inhibits cholinesterases, enhances the response of the muscle to acetylcholine, and d-tubocurarine blocks the response.[158] Acetylcholine causes a contraction of the body wall muscles of *Caenorhabditis elegans* and this is also blocked by tubocurarine.[212] Adrenaline, histamine, pilocarpine or strychnine, which have a pharmacological effect on nerve–muscle preparations of most animals, do not affect nerve–muscle preparations of *Ascaris*, but nicotine has the same effect as acetylcholine. The addition of acetylcholine to the bathing medium of a muscle preparation of *Ascaris* results in a marked reduction in the resting potential of the muscle. If it is released in the synaptic area between the finger-like processes of the innervation process and the nerve then depolarisation of the muscle occurs, but not if it is applied to small areas of the membrane enclosing the nucleated portions of the muscle or the innervation process. This indicates that the receptors for acetylcholine are restricted to this region of the muscle. Small amounts of acetylcholine are continuously released from the nerve cord at the junction with the innervation processes and this maintains the membrane potential of the muscle.[158]

Acetylcholine is broken down in the tissues by cholinesterases and these enzymes, including acetylcholinesterase, have been demonstrated to be present in several nematodes, but the amounts present vary enormously. The enzyme(s) are present in the innervation processes, the nerve–muscle junctions, the sense organs, the nervering and, possibly, the pseudocoelomic fluid.[8,158]

Large amounts of acetylcholinesterase are also present in the sub-ventral glands associated with the excretory system of several trichostrongyle nematodes (*Nippostrongylus, Trichostrongylus*) and in glands associated with the amphids of *Necator.* The enzyme in these glands is secreted to the exterior and may act on the sensory receptors or the muscle of the intestine of the host when the nematode attacks the mucosa. A similar phenomenon occurs in several biting animals (leeches, ticks, snakes) which secrete this enzyme in their saliva.[11,12,13,24]

Microelectrodes inserted into the non-contractile part of *Ascaris* muscle show that it has a resting potential of 20–30 mV and this is interrupted by rhythmical spike potentials, which originate at the nerve and are conducted along the innervation processes. The part of the muscle cell containing the myofilaments (contractile part of the cell) has a resting potential of 40–60 mV. Electrical stimulation of the ventral nerve cord gives rise to muscle potentials similar to spontaneous ones and these can be recorded in muscle cells up to 1 cm away from the stimulating electrode. The wave of excitation travels at about 60 m s^{-1}. Contraction of the muscle cells in *Ascaris* is brought about by rhythmic action potentials which occur at frequencies of up to twelve per second. The action potentials are myogenic in origin and arise from the membrane of the muscle innervation process at its junction with the nerve when depolarisation of this region occurs. The spike potential then spreads along the innervation process to the contractile part of the cell, resulting in contraction of the muscle.[9,158] The level of membrane potential and the frequency of depolarisation are controlled by neurohumors in the nerve cord. Two neurohumors appear to be liberated by the nerve axons of nematodes at their junction with the innervation process of the muscle. One, which is probably acetylcholine, results in depolarisation and increased frequency of firing; the other, which is probably γ-aminobutyric acid (GABA) causes hyperpolarisation of the membrane (figure 7.5) and brings about complete suppression of the rhythmic action potentials (it relaxes the body muscles of *Caenorhabditis elegans*[212]). These two neurohumors probably occur in excitatory and inhibitory synapses respectively.[158]

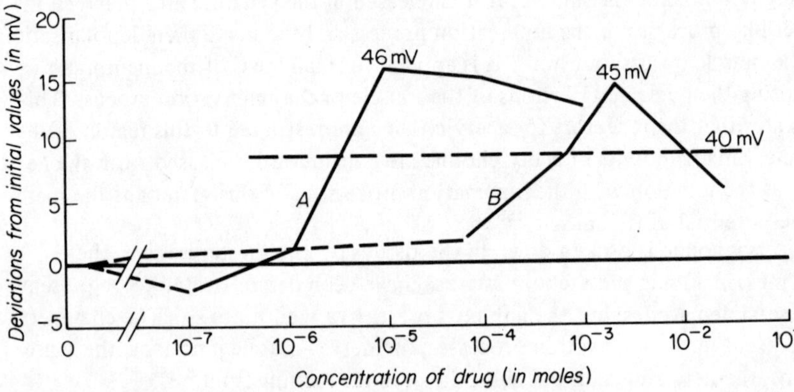

Figure 7.5 The effect of γ-aminobutyric acid (*A*) and piperazine (*B*) on the membrane potential of *Ascaris* muscle cells. Ordinates represent average membrane potential in mV, expressed as the difference from the initial values in the absence of drugs. Abscissae represent concentration of the drugs in moles. Note that the level of membrane potential induced by both drugs is almost equal (about 46 mV) but that the concentration of piperazine necessary to reach such hyperpolarisation is much greater than that of γ-aminobutyric acid. (After del Castillo.[158])

Certain catecholamines, notably epinephrine (adrenalin), norepinephrin and dopamine are important chemical transmitters in the nervous system of many animals. Juveniles of *Caenorhabditis elegans* apparently contain neither epinephrine nor norepinephrine but they contain significant amounts of dopamine (9×10^{-6} ng per individual). This catecholamine is the transmitter chemical in the cephalic and deirid neurons (about 1000 μg/g in the sensory endings) of juvenile and hermaphrodite adult *C. elegans* (section 9.1.2) and the male has six additional dopaminergic neurons in the tail. Similar dopaminergic neurons are present in the head region of juvenile and adult *Ascaris* but the organisation in the tail is rather different. Mutants of *C. elegans* which have reduced levels of dopamine are still sensitive to mechanical stimulation, probably because sensllla adjacent to the dopaminergic-deficient sensilla are still sensitive to touch. There is some evidence that mating efficiency is reduced in males which have dopamine-deficient neurons in the tail.[212]

Nematodes apparently possess both inhibitory and motor nerve fibres, as electrical stimulation (frequencies above five per second) in the region of the nerve ring of *Ascaris* causes inhibition of both spontaneous and electrically stimulated contractions of the muscles in the body wall. Inhibition also occurs in response to stimulation at frequencies shorter than 0.1 per millisecond, which implies that the effect is mediated by nerves. Cutting the ventral cord abolishes 80 per cent of the inhibitory effect on contractions of the musculature when stimuli are applied to the nerve ring. Cutting the dorsal cord reduces the inhibitory effect further.[158,189]

Recordings made of the potentials in two muscle cells, one of which is in the dorsal part of *Ascaris* and the other in the ventral, show that there is no correlation between the timing of the two sets of pulses. Recordings from muscle cells a few millimetres apart in the same section (dorsal or ventral) of the nematode show a correlation between the two records. In muscle cells at the same distance from the head but at different distances laterally from the longitudinal nerve, the muscle cell nearest the nerve gives the first pulse. If recordings are taken from muscle cells situated one in front of the other and equidistant from the nerve, then the anterior muscle cell gives the first pulse. A transverse cut between these two muscle cells does not destroy the correlation between the pulses of the two cells. If the nerve lying parallel to these two cells is cut between them, the correlation is abolished and independent pulses are recorded from the two cells. This indicates that the longitudinal nerves in the dorsal and ventral lines co-ordinate the activity of the muscles in the dorsal and ventral parts of the nematode respectively.[9,158]

Nerve–muscle preparations from juvenile *Phocanema decipens* have been used to study the response to electrical stimulation and the occurrence of spontaneous rhythmical activity when the temperature is lowered from 20 °C to 0 °C. Anodal stimulation brings about contraction at all temperatures below 20 °C and 0 °C, but rhythmical activity ceases as the temperature is lowered. Cathodal stimulation causes relaxation at the higher temperatures, but as the temperature is lowered it causes no response and then it brings about contraction at the lower temperatures. The temperature at which relaxation ceases has no relation to the temperature at which rhythmical activity ceases. This is further evidence that relaxation and rhythmical movements are the result of two independent processes.[9]

In summary, it is thought that during locomotion in nematodes, the main role of the nervous system is a modulatory one. The nervous system superimposes co-ordination and control of the level of membrane potential and frequency of depolarisation on the inherent activity of the muscle cells and this is brought about by both excitatory and inhibitory fibres. The muscle cell of nematodes appears to have some of the properties of a proprio-receptor, but no structure resembling a proprio-receptor has been detected in these muscles.

7.3 Pharmacology

The effects of acetylcholine, nicotine, d-tubocurarine, eserine, γ-aminobutyric acid, epinephrine (adrenaline), histamine, pilocarpine and strychnine on nematodes, notably *Ascaris,* have been described in section 7.2.4, but none of these is used as anthelmintics nor as nematicides. However, several chemicals which are used as nematicides or as anthelmintics are known to affect the neuromuscular system of nematodes. Santonine (5-hydroxytryptamine) interferes with the coordinating nerve impulses that originate at the nerve ring in *Ascaris* and results in paralysis of the worm.[158] Bephinium salts, which are used against nematodes in the small intestine of vertebrates, are similar to acetylcholine in their action. The bephenium ion bears a strong structural resemblance to acetylcholine but is five times more active in causing contraction of *Ascaris* muscle; vertebrate muscles are relatively insensitive to these bephenium salts. Methyridine is another anthelmintic which brings about a neuromuscular block by means of a depolarising effect. Paralysis of *Ascaris* nerve–muscle preparations is reversed by d-tubocurarine and by piperazine, but not by acetylcholine. Methyridine also affects the muscles of vertebrates but nematode tissues are much more sensitive to this inhibition and the therapeutic value of this anthelmintic appears to depend on this difference in sensitivity. All of these anthelmintics apparently affect the nematode by inactivating the depolarising receptors in the worm.[159] Tetramisole inhibits the fumarate reductase mechanism in several nematodes (section 4.5.2). However, contraction of the muscles of *Ascaris* occurs at lower concentrations than those needed to bring about complete inhibition of ATP synthesis by the fumarate reductase pathway. Apparently the initial effect of tetramisole is on the muscles of the nematode, where it reduces the membrane potential, resulting in contraction of the muscle. This activation of the muscle breaks down the myosin-bound ATP to ADP. Rephosphorylation of the ADP is greatly reduced by the inhibiting effect of the drug on the fumarate reductase pathway, leading to exhaustion of ATP supplies in the muscle and to irreversible block of the contractile elements in the cell.[160] Motility of *Litomosoides, Dipetalonema* and *Brugia* ceases within five minutes in low concentrations of tetramisole and it is thought that the primary effect of the drug is on the neuromuscular system of these filarial nematodes.[121]

A number of organophosphorus and carbamate compounds are important insecticides and they apparently affect the insect by inactivating cholinesterases in the animal. Some of these have been developed as anthelmintics or as nematicides. Haloxon is an organophosphorus compound which is used in veterinary medicine. It inactivates the cholinesterases of several animal-parasitic nematodes (for example, *Haemonchus contortus*) and this reaction is irreversible. It also affects sheep erythrocyte cholinesterase but the enzyme reactivates spontaneously and this probably explains why the anthelmintic is relatively non-toxic to the host but is toxic to the nematode. Not all nematodes are affected, however, as *Bunostomum trigonocephalum* and *Dictyocaulus filaria* have cholinesterases which resemble the mammalian enzyme in that they reactivate after inhibition with haloxon. A number of organophosphorus compounds have been tested for their activity against the acetylcholinesterases of *Nippostrongylus brasiliensis* and of bovine erythrocytes and the nematode enzymes are at least ten times more sensitive than the mammalian enzyme to inhibition by these compounds.[159] Aldicarb is a carbamate that rapidly inhibits body activity of juvenile *Heterodera* and enzymic hydrolysis of acetylcholine in *Panagrellus*. It is thought to work as a nematicide by inhibiting the cholinesterases of the nematode, thus bringing about paralysis. Phorate (an insecticide) is another carbamate which is active against certain nematodes and also seems to act against the cholinesterases of the worm. These substances do not always kill plant-parasitic and free-living nematodes at practical application rates, but paralyse them, and the nematodes often recover if the chemicals are removed from the environment.[213,214,215]

A number of chemicals activate the hyperpolarising receptors of nematode nerve–muscle junctions and some are important anthelmintics. Piperazine salts, which are widely used in the treatment of humans infected with *Ascaris* and *Enterobius*, increase the resting potential of *Ascaris* muscle from 30 mV to about 45 mV (figure 7.6) and bring about flaccid paralysis of the muscle. Rhythmic action potentials, which initiate contraction of the muscle, disappear when the resting potential rises above 40 mV. This hyperpolarising action takes place at the nerve–muscle junction, which is where the piperazine receptors overlap with the acetylcholine receptors. The changes that it produces are similar to those produced when the inhibitory nerves have been stimulated by GABA (figure 7.5) and are thought to be due to activation of the hyperpolarising neuromuscular receptors. This lowers the response of *Ascaris* muscle to acetylcholine (figure 7.6).

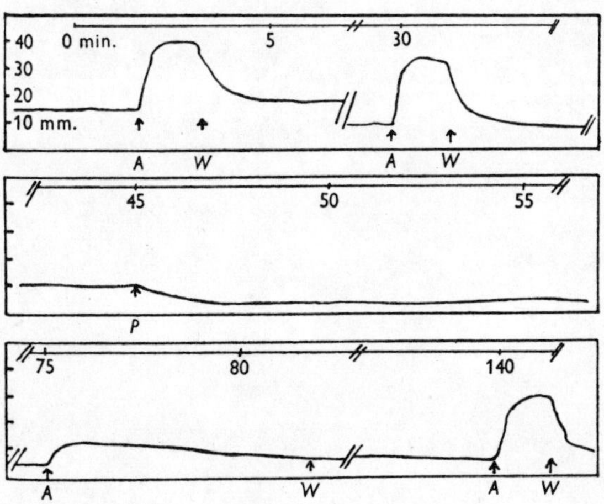

Figure 7.6 Tracing from Sanborn recording of a split preparation of *Ascaris* showing the response of the preparation to acetylcholine (*A*) (1 μg/ml), to washing (*W*), and the effect of piperazine (*P*) (250 μg/ml) on the response to acetylcholine. [After Norton and De Beer, *Amer. J. Trop. Med. Hyg.*, 6 (1957).]

Activation of the inhibitory receptor by the drug increases the permeability of the membranes at the neuromuscular junction to chloride ions. Piperazine also brings about a decrease in succinate formation (section 4.5).[158,159]

The action of piperazine is reversible, as paralysed worms will recover motility if transferred to saline at 37 °C (figure 7.6), but paralysis lasts long enough for the worms to be swept out of the host. This drug has little or no effect on the neuromuscular junctions of the mammalian host. A number of piperazine derivatives are used as anthelmintics and it is thought that their action is similar to that of piperazine.[158,159]

Several other anthelmintics appear to affect the neuromuscular system of nematodes but the mechanism of action of many is not known.

8 Locomotion

8.1 The Hydrostatic Skeleton

Nematodes differ from most other triploblastic animals in that they do not have circular muscles in their body wall. The longitudinal muscles are usually arranged in two dorsal and two ventral fields but in most species they function as one dorsal and one ventral field. Contraction of any region of one field normally causes a flexure of the animal rather than a change in diameter or a corrugation of the cuticle. These deformations are opposed by the combined effect of an anisometric cuticle and a high hydrostatic pressure within the pseudocoelom. Changes in pressure in the pseudocoelom are transmitted in all directions and restore both body shape and the length of the muscles when they relax. Furthermore, contraction of one muscle tends to oppose all others by altering either their length or tone.

An elastic cuticle, containing a mesh of fibres, and the free movement of fluid throughout the pseudocoelom are important features in the locomotion of *Ascaris.* In this nematode three series of many parallel fibres form cylindrical helices in the cuticle and one helix rotates in the opposite direction to the other two. The opposing fibres cross and form a mesh of minute parallelograms (figure 8.1). The fibres are inextensible, but a change in shape of the cylinder they enclose is possible by drawing the helix apart or by reducing its length. Such a system has the important feature that this causes a change in the enclosed volume; alterations in diameter do not exactly compensate for changes in length. The fibres in *Ascaris* are so arranged that the contraction of the longitudinal muscles attempts to reduce both the body length and the volume of the animal (figure 8.1). This cannot readily occur because the nematode does not leak and the body contents, including the pseudocoelomic fluid, are incompressible. A change in diameter is resisted and consequently contraction of one field of longitudinal muscle increases the hydrostatic pressure within the body contents and the shortening muscle cells cause a flexure of the body.[2] The pseudocoelomic fluid of *Ascaris* has a hydrostatic pressure of approximately 70 mm Hg but large and often rhythmical variations between 16 and 225 mm Hg occur during locomotion. Undulatory waves are produced by alternative waves of contraction in the two fields of longitudinal muscles. Their functional opposition is effective because it is mediated by the anisometric fibre system in the cuticle and the pseudocoelomic fluid which acts as a hydrostatic skeleton.[2]

The similarity of form amongst nematodes indicates that many use a method of co-ordination mechanically similar to that of *Ascaris.*[2,9] However, small nematodes often have a relatively smaller volume of pseudocoelomic fluid than larger species and may transmit much of the hydraulic force through the partly deformable body tissues and the fluid gut contents.[217] Many animal-parasitic nematodes

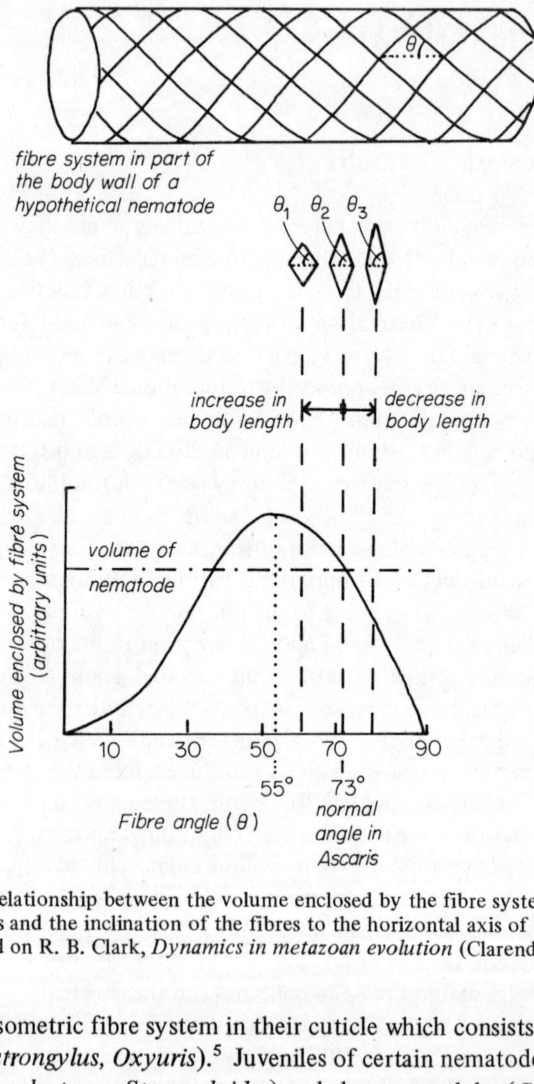

Figure 8.1 The relationship between the volume enclosed by the fibre system found in many nematodes and the inclination of the fibres to the horizontal axis of the animal (angle θ). [Based on R. B. Clark, *Dynamics in metazoan evolution* (Clarendon Press, Oxford 1964).]

possess an anisometric fibre system in their cuticle which consists of two fibre layers (*Nippostrongylus, Oxyuris*).[5] Juveniles of certain nematodes (*Nippostrongylus, Ancylostoma, Strongyloides*) and also some adults (*Capillaria, Hirschmaniella, Panagrellus, Ditylenchus*) contain a striated layer in the cuticle which may provide some of the tensile strength that is required of the cuticle during locomotion[6] (section 2.1.1). The second-stage juvenile of *Heterodera rostochiensis* is an example of a nematode in which the striated layer is present but the fibre layer is apparently lacking. The striations may behave as struts lying perpendicular to the surface of the cuticle between more elastic material. Such an arrangement would allow flexibility of the cuticle during locomotion but may provide the limit to its elasticity that is essential for muscular contraction to result in body flexure.[26]

The system of locomotion in nematodes is of paramount significance in our understanding of their organisation. The pseudocoelomates lack a splanchnic muscle system and the nematode cannot, therefore, prevent a collapse of its gut by the high hydrostatic pressure. The pharynx is muscular and pumps food into the intestine against this hydrostatic pressure. Also, structures in the pharynx and pre-rectum act as valves which prevent the involuntary loss of gut contents from the animal; during defecation material is forcibly ejected by the high internal pressure inside the nematode. In some species a muscle network surrounds the posterior end of the intestine and assists in the ejection of the faeces.[10] It has also been suggested that the structure of the reproductive system and the absence of motile flagella from spermatozoa or the compressed internal organs may also be related to the effects of this pressure.[2,9] This may be a valid generalisation but it should be noted that flagella are present in the intestine of the free-living nematode *Eudorylaimus* (section 1.1).

8.2 Sinusoidal Movement

During sinusoidal movement, alternating phases of contraction pass posteriorly along the two muscle fields and this produces waves which travel along the length of the nematode. They are produced by the dorsal and ventral fields of muscle, and body waves are therefore formed dorso-ventrally. Consequently, nematodes are unusual because they lie on their sides and not their ventral surface while moving over a substrate. The forces involved in transforming some of the work done during undulation into a forward progression can be described in a simple but approximate manner. In an ideal, rigid environment, such as occasionally occurs in soils, the animal forms symmetrical bending waves with all parts of the body following a sinusoidal track of one body width. The travelling body waves apply a continuous, steady force against a series of rigid particles lying on each side of the animal. The reaction to this force exerted by the relatively immovable particles can be resolved into two components acting at right angles, but that in the direction of travel is particularly important. It increases with wave amplitude and forward movement occurs when frictional forces are overcome (figure 8.2). Locomotion under forces which are approximately similar to those described may occur in some environments but undulatory propulsion often involves more complex forces than those described in this simplified ideal.[218,220]

8.2.1 Swimming

Some nematodes move through fluids without contact with a substrate. Analysis of swimming by a small animal is extremely complex but certain factors are of major significance. During swimming the nematode generates waves at a faster rate than its speed forward because, unlike locomotion between a series

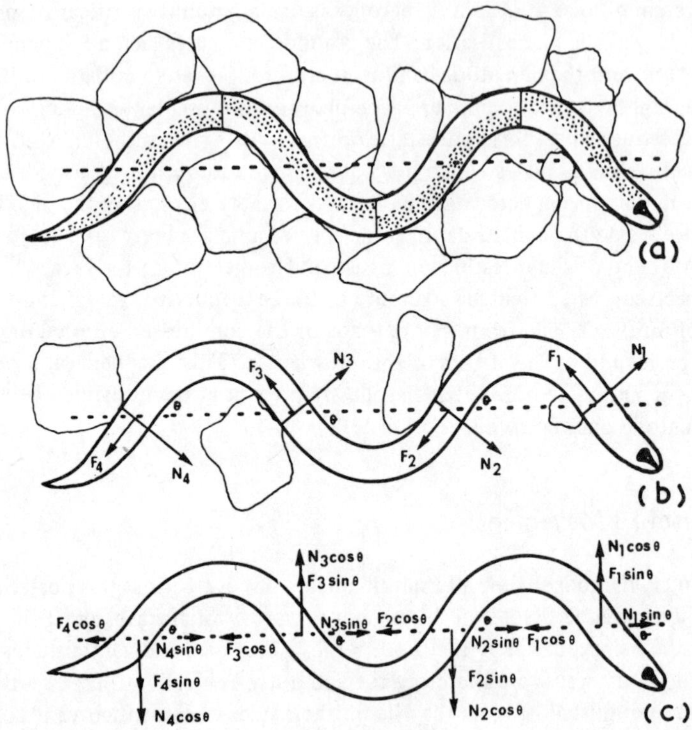

Figure 8.2 (*a*) A nematode moving along a sinusoidal track among soil particles. There is a progressive change of curvature along its length, each half-wave or S-curve operating as a single propulsive unit. The dotted half of each sector indicates that tension is developed in the muscles on that side. (*b*) The normal reactions (N_1–N_4) between nematode and soil particles induced by the backward pressure of the nematode. The friction forces F_1–F_4 act tangentially along the nematode. The angle between the nematode and the main axis of progression is θ. (*c*) Each of the normal reactions (N_1–N_4) and friction forces (F_1–F_4) is resolved along axes parallel and normal to the main axis or progression. Forward movement occurs when the amplitude of the body wave indicated by θ is sufficient for $\Sigma N_1 \sin\theta \equiv \Sigma F_1 \cos\theta$. (After Wallace.[218])

of immobile particles, a part of the energy of each wave is dissipated in slip of the body relative to the surrounding medium. The effect is most marked at low viscosities and few nematodes swim well in water. Nematodes that are able to move quickly through water (*Turbatrix, Monhystera*) are elongate and rapidly propagate sinusoidal waves that tend to increase in amplitude and wavelength towards the tail[219] (figure 8.3). A lower specific gravity than most soil nematodes, together with a centre of gravity within the posterior half of the animal, are also adaptations to swimming in *Turbatrix* and enable this animal to remain bouyant in water. The problems associated with swimming are altered by an increase in the viscosity of the medium. The efficiency of swimming is greater as there tends to be less slip of the body waves but the forward velocity may be reduced unless the nematode is able to raise its power output with an increase in the viscosity of its surroundings. Many nematodes can swim in viscous media and waves with a lower frequency and shorter wavelengths are successful in these conditions. A large number of species occur in media such as mucus, faeces, some muds and decaying matter, all of which are suitably viscous fluids for swimming in this manner.

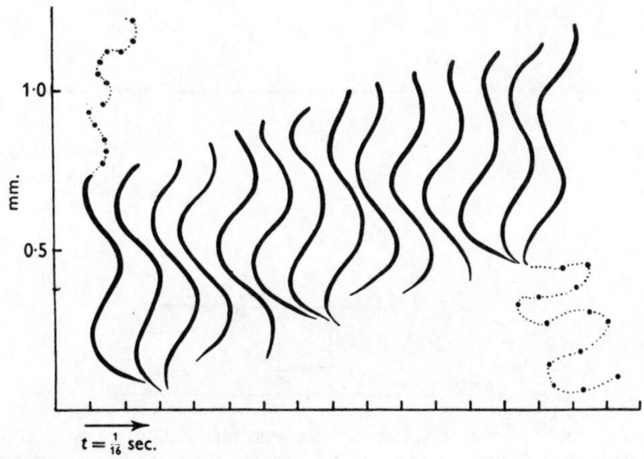

Figure 8.3 Successive positions at intervals of $\frac{1}{16}$ second of *Turbatrix aceti* moving in deep water. The amplitude of transverse movements of the tail is greater than that of the head. (After Gray and Lissman.[219])

8.2.2 Movement in Water Films (Crawling)

Although nematodes are aquatic animals, surprisingly few are capable of swimming effectively in deep water. Many move more rapidly in continuous water films or through scattered drops of water in contact with the surfaces of soil particles, organic matter and plants. Movement in water films is associated with waves showing a lower rate of travel along the body than occurs in water. This may well indicate that a different type of co-ordination is involved (section 9.2). Additional forces resisting slip are available to the animal in water films. They are due to surface tension forces and only occur when the film is thinner than the diameter of the individual, resulting in contact between the animal and the surface of the water. The water envelopes the animal, consequently the film is deformed and this produces a vertical component of the surface tension force pressing the nematode against the substrate. Thinner films suffer greater distortion and so a greater vertical component results than in those films that are just less than the diameter of the nematode (figure 8.4). In thicker films some lateral slip may occur, causing an asymmetrical distortion of the film (figure 8.4) which produces a horizontal component opposing slip. Eventually, further lateral

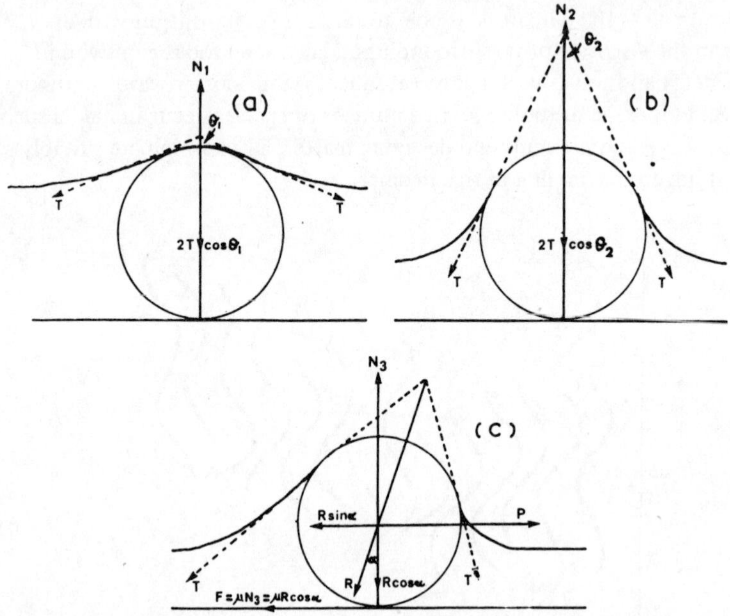

Figure 8.4 The external forces acting on a nematode at rest and during motion in a water film. The nematodes are drawn in transverse section in: (*a*) a thick film of water; (*b*) a thin film of water; and (*c*) during motion when the water film is displaced. *T* is the force of surface tensions, *N* the normal reaction between nematode and substratum, *R* is the resultant of the surface tension force in (*c*), *F* the friction force, μ the coefficient of friction and *P* the nematode's propulsive force. (After Wallace.[218])

movement is prevented by this progressively greater force and there is a useful contribution of the body wave to forward progression. In thinner films surface tension forces allow a firmer grip on the substrate and lateral slip of the body is reduced. No slip occurs in films of ideal thickness and all parts of the animal follow a sinusoidal track of one body width as in the ideal rigid environment described earlier. In very thin films of water the vertical component of surface tension is even greater and the nematode is held more firmly against the substrate. In these conditions the nematode tends to produce undulations with short wavelengths but even so forward movement is greatly reduced[198,218] (figure 8.5). Nematodes may also have a limited ability to control their speed forward in films and between particles. Experiments with *Meloidogyne* juveniles suggest that an increased effort is possible by increasing the proportion of muscle cells producing body flexure.[221] Prevention of movement in dry conditions is partly due to the increased vertical component of surface tension but nematodes also lose water readily if the relative humidity falls. Any loss of water will reduce the hydrostatic pressure of the animal and will tend to interfere with the efficient working of the locomotory system; eventually the dry conditions may desiccate the animal (section 5.1.4).

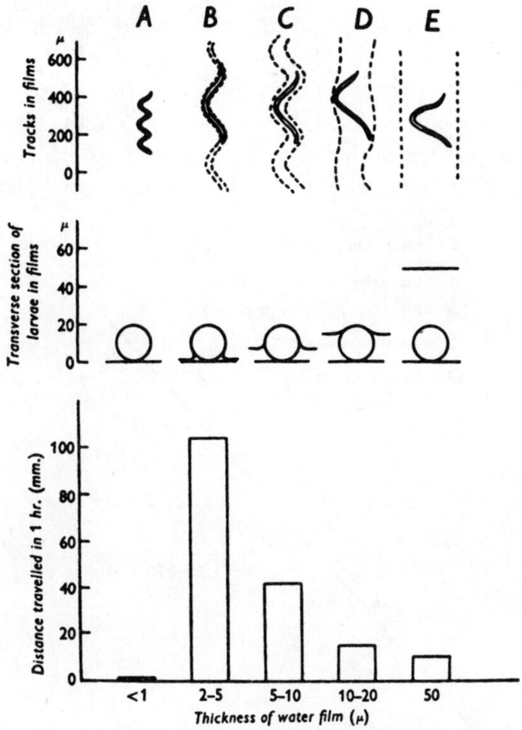

Figure 8.5 Movement of juveniles of *Heterodera schachtii* in different thicknesses of water film on glass showing the relationship between the thickness of the film and nematode mobility. [After Wallace, *Ann. appl. Biol.* (1958).]

Nematodes that occur in thick water films include ectoparasites, which attack the aerial parts of plants and some animal-parasitic juveniles that migrate up herbage prior to ingestion by a host. Many are moderate swimmers (figure 8.6) but move rapidly even in thick water films. Damp conditions that are suitable for movement are not continuously available to these animals and many show a range of adaptations that ensure success in this variable habitat. Some species can withstand desiccation by becoming cryptobiotic (*Tylenchus, Anguina, Haemonchus*; section 5.1.4). *Aphelenchoides ritzemabosi* moves extremely rapidly over wet chrysanthemum leaves and is particularly favoured on this plant by the presence of epidermal hairs between which it glides.[198] Individuals can feed as ectoparasites on chrysanthemum but tend to enter the leaf through stomata and become endoparasites during drier conditions. The animal also shows well co-ordinated behaviour on leaves and this may reduce the time taken to locate stomata to a minimum during short periods of damp conditions (section 9.3). Infective juveniles of many animal-parasitic Trichostrongylidae also migrate up stems or leaves in moist conditions. The animals hatch in host faeces but later move onto the surrounding soil where they are apparently attracted to grass roots. Many show a level of activity in water and water films comparable to that of plant-leaf parasites (figure 8.6). The juveniles tend to continue movement until an interruption occurs in the substrate. This may aid their upward movement because the shape of grass blades offers a greater mean free path in the vertical rather than the horizontal plane.[222]

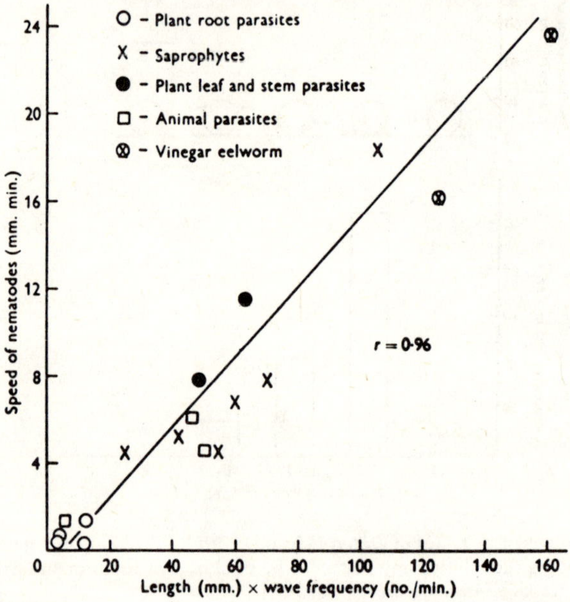

Figure 8.6 The relationship between the speed of nematodes and the product of their length and wave frequency when moving in deep water. [After Wallace and Doncaster, *Parasitology* (1964).]

The tendency to ascend herbage has a considerable value to these nematodes because it increases the chance of ingestion by a herbivorous host and this is essential for their further development.

Clearly, movement over plants can be a hazardous event for nematodes but most are able to move rapidly whenever suitable conditions are available. At other times the animals remain in the soil or in moisture held between bud and leaf axils and some can survive desiccation if exposed to dry conditions.

8.2.3 Movement in Soils

The movement of nematodes in soil is particularly influenced by the physical structure of the soil and by its water content. These two factors are closely interrelated but it is simpler to consider their effects separately. Nematodes are too small to displace many particles in most soils and are therefore restricted to the pores formed between the packed particles. Experiments with artificial soils made of graded sand grains have shown that nematodes move most rapidly when their diameter is slightly less than that of the pore necks. In these conditions the nematode is able to bend freely through the pores and around sufficient rigid particles for efficient sinusoidal movement to occur (figure 8.2). Pores of greater width than this optimum size may allow a significant lateral slip of the body and so reduce the speed of forward movement. Channels narrower than the body diameter act as barriers to movement because most nematodes seem unable to disturb surrounding particles by burrowing. The size of the pore relative to the body diameter is particularly important, consequently maximum mobility occurs for second-stage *Heterodera* juveniles in pores which are smaller than those required by the larger, fourth-stage juvenile of *Ditylenchus.*[167,198,218]

Natural soils contain a wider range of particle sizes than a graded soil. A number of channels ramify from any point in a soil but their diameter will tend to vary along their length. At times the pore will be too narrow for a nematode to pass. The theoretical distance that a nematode can travel through the soil to this position is called the free path and its mean length available for the nematode in any direction will decline with an increasing proportion of smaller particles. Nematode movement in soils may be co-ordinated in a manner that permits the animal to select a new direction of travel when a particular channel becomes too narrow (section 9.2.1). In general, soils with a high clay content reduce mobility more than do sandy soils, except where numerous cracks and fissures give alternative pathways for movement. Soil-inhabiting nematodes, such as *Longidorus* and *Xiphinema*, may be restricted by their large size to a small proportion of the soil volume, perhaps as little as a few per cent of the soil pore space in their habitat. It is probable that these nematodes are particularly dependent for movement on channels formed in the soil by plant root systems.[223,224] Soil-inhabiting nematodes as large as *Longidorus* are unusual and the small size of most species may be correlated with the ideal length and breadth appropriate for movement in the channels that lie between soil particles.[225]

The second important factor that influences the movements of nematodes in soil is the amount of water present in the pore spaces. Soil-water content increases after rainfall or irrigation and falls later due to drainage and evaporation. Each pore drains when the drying processes exert a force equivalent to the suction necessary to draw water from it against a surface tension force. The surface tension force resisting loss of water acts around the pore neck and is inversely proportional to the neck diameter. This relationship between the suction applied to a soil and its water content can be determined experimentally and is known as the moisture characteristic of the soil. A greater suction is required to empty a soil with many small pores than is necessary for one with mainly larger particles. The shape of the moisture characteristic curve of any soil is, therefore, related to soil type; sandy soils lose water more readily than those with a higher silt or clay content. The changes in soil moisture influence the thickness of films and the presence of water lenses available to the nematode during movement. In a typical sandy soil, nematodes move most readily when the pores are just emptying (figure 8.7). In this situation the surface tension forces between the nematode and its substrate are sufficient to reduce lateral slip. As the soil dries, further movement is probably restricted by the greater component of surface tension force holding the nematode to the substrate and by the reduced availability of free water. On the other hand, rapid movement in waterlogged soils is dependent not only on swimming ability but also on an adequate oxygen supply. Sandy soils show a very well-defined relationship between the moisture characteristics and mobility but the correlation is more complex in some soils, particularly those that contain peat; this suggests that other factors can also be important.[167,198,218]

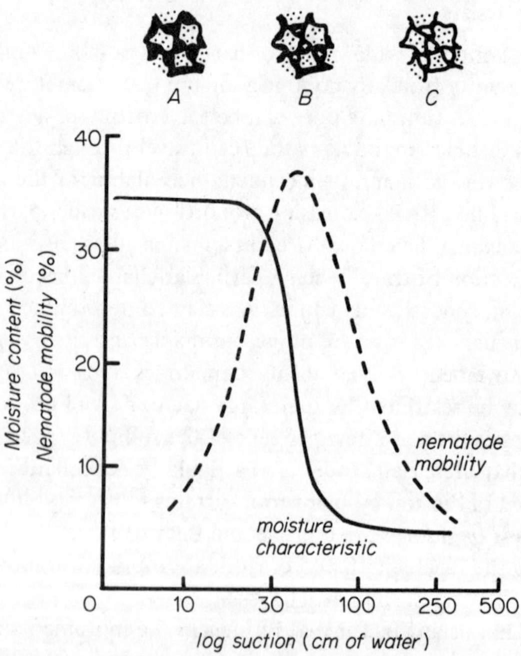

Figure 8.7 Mobility of a soil nematode related to the moisture characteristic of a hypothetical, sandy soil sample offering many pores suitable for sinusoidal movement. *A.* Pores full of water; *B.* pores emptying; *C.* pores empty. (Based on Wallace.[198])

Sufficient water for movement will always be available to the nematode just above the water table in the soil, but maximum pore size filled with water decreases with height above this level. Changes in climate on a diurnal or seasonal basis alter the moisture content of the soil above the water table and these effects are most marked in the uppermost soil layers. *Trichodorus* is a good example of a nematode that is influenced by changes in soil moisture. Prolonged wet weather in spring raises the water table and enables populations of this nematode to damage the root systems of young sugar beet plants in certain British sandy soils.

In a dry spring the seedlings are still capable of growth but escape damage by *Trichodorus*. In these conditions the lower moisture content of the upper layers of the soil restrict activity of this nematode to regions that are deeper than those occupied by the developing root systems of the sugar beet plants.[224]

8.2.4 Movement Within Animals

Penetration of infective juveniles of *Nippostrongylus* and *Ancylostoma* through the skin of mammals is essentially a mechanical process. Juveniles of *Nippostrongylus* frequently locate crevices that increase the purchase available to the nematodes as they push the stratified keratin apart. Both *Nippostrongylus* and *Ancylostoma* are able to penetrate from the outer surface of the skin to the stratum corneum in a few minutes and are no longer susceptible to desiccation after their initial entry. The juveniles move horizontally in the skin, particularly in the interface between the stratum corneum and the stratum granulosum, in a position that may offer the least resistance to movement through the skin. Passage through the dermis is a slower process than the initial entry. The dermis is sufficiently fluid for locomotion of *Ancylostoma* in this layer to be accompanied by appreciable lateral slip of the body waves. Purchase during entry into this layer is provided by that portion of the nematode still within the more rigid stratum corneum. Penetration of both *Ancylostoma* and *Nippostrongylus* involves the separation of the constituent cells of each tissue reached, but enzymatic secretions are not thought to play a major role in this process.[77,78]

Parasitic nematodes migrating through the tissues of their hosts may move freely between the cells in the manner described for *Ancylostoma* in the dermis of the skin. Intestinal nematodes are also able to move along the gut but most smaller species (*Nippostrongylus, Trichostrongylus, Nematodirus*) usually remain in close contact with the mucosa. Fourth-stage juveniles and adults of *Nippostrongylus* move back to their original position in the mucosa after displacement along its length.[245] *Nippostrongylus* is able to move by either two- or three-dimensional movement in viscous media or between solid particles (section 8.3). A single spiral wave may be used *in vivo* to brace the body between the villi and force all parts of the animal through this stationary loop (figure 8.8). *Trichostrongylus* may not move solely between the villi because it is also able to tunnel beneath the epithelial cells of the intestine.[149]

Ascaris is too large a nematode to maintain the close contact with the mucosa that is typical of many smaller species. It may swim in the viscous contents of the intestine and also brace itself against the wall of this organ, which will prevent its displacement by the normal movement of the gut.

8.2.5 Movement Within the Egg

The juveniles of many plant-parasitic nematodes form figure-of-eight curves within the egg. It has been suggested that this arrangement favours the propul-

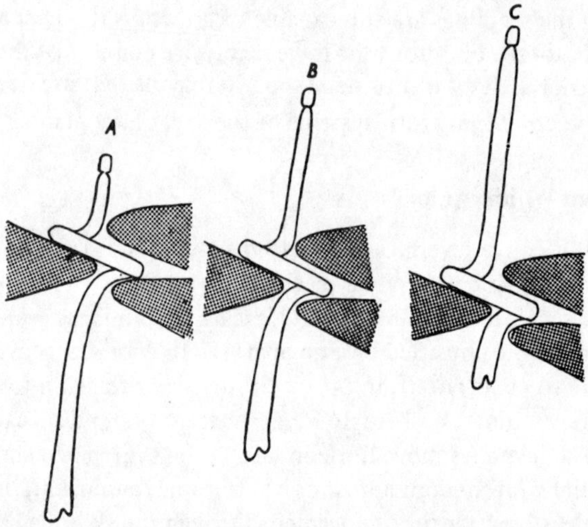

Figure 8.8 The locomotion of an adult *Nippostrongylus* removed from the intestine of the host and placed amongst moist sand grains. It is probable that similar movements are performed by the nematode amongst the villi of the host intestine. (After Lee.[226])

sive force required during movement by the unhatched juvenile. This configuration allows contraction of both main muscle fields, whereas a simple coil without torsion would be unsuitable for movement as only one field of musculature would have the opportunity to contract. Movement around the egg and the thrust associated with emergence may also be favoured by the non-spherical shape that is typical of the eggs of many species. A decrease in the radius of curvature at the pole of the egg will be associated with a greater flexure of the body and may increase the thrust produced by the contracting musculature.[225]

8.3 Other Forms of Movement

There is some evidence that nematodes can produce sinusoidal waves differing in speed and direction of propogation. Nematodes can travel backwards and unusual wave forms may also be associated with specialised activities during feeding.[9] Many species produce complex flexures whilst searching for food near the mouth. Some nematodes can form a helix either with or without body torsion. The helix without torsion is reminiscent of normal sinusoidal movement (figure 8.9*a*) except that not all the muscle blocks are relaxed in the phase between contraction of either the dorsal or ventral fields of the animal (figure 8.9*b*). Instead, the portions of the body lying across the axis of progression show a contraction. One dorsal and one ventral quadrant are contracted as a pair in the leading half of the body wave, with the opposite pair contracting in a trailing portion of each wave. Co-ordination requires a sequential wave of contraction down single quadrants, rather than in dorsal or ventral pairs as occurs in sinusoidal movements (section

8.2). The result is that the nematode can move forwards by wrapping each surface in turn around a central rod;[218] adult female *Mermis* probably move up plant stems to lay eggs in this way.

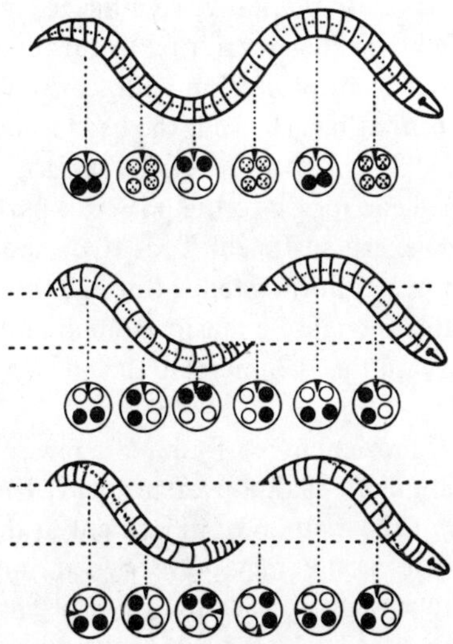

Figure 8.9 Undulatory movement and muscular activity in (*a*) a single plane, (*b*) spiral movement without torque, and (*c*) spiral movement with torque. The transverse section of the body with its dorsal and ventral pairs of muscles is shown diagramatically at different places along the body as viewed from the posterior end of the nematode. The small circles denote muscles, those contracted are coloured black, partially contracted are stippled, and relaxed are white. To indicate a spiral form, the nematodes in *b* and *c* are drawn around an imaginary cylinder. The black triangle in each transverse section denotes the dorsal side of the nematode, and the dotted line along the length of each nematode represents the right lateral line. (After Wallace.[218])

The second form of helix with one surface always uppermost involves torsion and is infrequently produced by vermiform animals (figure 8.9*c*). A continual winding of the body around solid cylinders longer than one body length is not possible in this way, for the complete body helix involves the contraction of only one pair of muscle quadrants throughout the length of the nematode; once formed the helix must be unwound or the animal is locked in position. However, forward progression is possible under certain circumstances. It can be seen from the longitudinal ridges in the cuticle of *Nippostrongylus* that this nematode can form spirals without torsion. This nematode changes to body waves of this type in viscous media. Movement is possible if single loops of this kind move backwards along the animal. This sytem of co-ordination is probably used to move between projections such as intestinal villi. One or two loops brace the body of the animal between a few villi (figure 8.8) and all parts of the body pass forward through the loop.[226] Continual forward locomotion requires the sequential formation of single loops by the anterior region of the animal. Many nematodes are capable of producing variable wave forms in the anterior region. Helices of both forms are probably produced frequently and they indicate the ability to co-ordinate muscle quadrants individually.

The infective juvenile stages of *Neoaplectana carpocapsae* show an interesting form of movement that has clear adaptative significance in the location of its insect hosts. The juvenile waves the anterior part of its body free of the substrate and may move onto any adjacent projection with which it makes contact. If an attachment of this kind cannot be achieved, the head is looped around to near the tail. The animal is held in this position by the surface tension forces of a moisture film binding the head region to the posterior part of the animal. Eventually, the nematode exerts sufficient force to overcome this restraint and its release propels the animal a short distance through the air.[227] The surface tension forces probably enable the muscle tension to build up within the body wall before acceleration starts and this will increase the mean acceleration and final velocity of the nematode.

More unusual forms of movement are found in a few species. *Criconemoides* has a series of cuticular annulations that overlap to give backward-facing projections. Little is known of the method of movement of these nematodes in soil but a phase of muscular contraction may pass forward causing the annulations to telescope as it passes. The cuticular projections may be able to slip forward at this time but otherwise prevent backward slip of the body. Each completed phase of contraction would result in a slight forward movement.[228] The cuticular projections may also aid in bracing the animal while it penetrates roots with its exceptionally large stomatostylet. *Desmoscolex* is another nematode which undergoes changes in body shape during progression. The animal has alternate annulations of thickened and flexible cuticle. Waves of contraction pass forward along the nematode and draw the thicker annulations together. Backward-facing bristles on the thicker annulations oppose any tendency of the body to slip posteriorly and thus the nematode moves forwards. *Chaetosoma* may move somewhat differently, by looping. The animal first anchors its head to seaweed by means of an adhesive secretion and then draws up its hind region. Bristles at the posterior end then adhere to the substrate while the head is released and extended further forward.[229] The movement of *Criconemoides, Desmoscolex* and *Chaetosoma* may require a specialised arrangement of the nervous system, although co-ordination may also be achieved by a hydrostatic skeleton and the mechanical stretching of adjacent muscles. Further work on these aberrant forms of movement is required.

8.4 Dispersal Mechanisms

Unlike many insects, most nematodes have extremely limited powers of active dispersal. Active migration from soil to plant roots or from faeces to herbage is an important event in the life history of many plant- and animal-parasitic nematodes, but the distances travelled are usually measurable in centimetres not metres. The spread of a population within a habitat is, however, often aided by at least one form of passive dispersal.

Surface drainage after rainfall is a particularly important factor in the dispersal of many nematodes. This effect can readily be seen in fields of lucerne where the distribution of stunted plants infected by *Ditylenchus dipsaci* spreads much more rapidly down a slope from a restricted foci of infection than uphill. Nematodes may also percolate through soil with the downward movement of water following heavy rainfall.[202] This may not be a totally passive process. Active individuals apparently facilitate their movement through the soil and tend to spread more rapidly than inactive species of a similar size.[198]

Infective juveniles of *Dictyocaulus viviparus* (a parasite of cattle) do not readily migrate, yet they become widely dispersed on herbage surrounding the faeces in which they develop. Their dispersal is brought about by the fungus *Pilobolus.* The infective stage of the nematode moves to the surface of the faeces, which is also heavily colonised by the fungus. *Pilobolus* discharges its sporangia violently in response to changes in illumination, catapulting those juveniles that have climbed up the sporangiophores onto adjacent pasture. Transport by this means is particularly important to the nematode as it must be ingested by a host for further development and cattle avoid grazing near their own faeces.[230]

Various agricultural activities favour the dispersal of plant-parasitic nematodes. Soil adhering to footwear, machinery, tubers, and all kinds of transplants, augment the dispersal of cysts of *Heterodera* by natural agents such as drainage and high winds. *Tylenchulus semipenetrans* on citrus roots and *Radopholus similis* in banana 'sets' are widely distributed as a result of the introduction of host plants to new regions. Some nematodes are capable of withstanding air-drying and become seed-borne, allowing them to be dispersed with seed and infect further host crops in subsequent seasons. Examples include *Anguina tritici* in wheat grain, *Aphelenchoides besseyi* on rice and *Ditylenchus dipsaci* on field bean and lucerne seed. *Ditylenchus dipsaci* and *Aphelenchoides ritzemabosi* can also resist desiccation within dried plant material and are, therefore, readily dispersed in fields and to cultivated plots by farm activities and wind action. Eggs within cysts of *Heterodera rostochiensis* can also withstand air-drying and this has been an important factor in the spread of this nematode. Cysts adhering to the surface of seed tubers were probably responsible for transport of this pest from the original sources of the parasite and of the potato in South America to most of the main potato-growing regions of the world. Its introduction, followed by spread to many parts of Europe, probably took seventy to a hundred years.[202]

There are many species of nematode that are associates of insects and they are often transported to new, transient habitats attached to these arthropods. The dauer juvenile of some nematodes that live in faeces are carried from one habitat to the next under the elytra of dung beetles or on the bristles of insects that alight on the faeces.[231]

In general, geographical isolation of parasites of higher animals from suitable host populations may occur less frequently than for plant-parasitic nematodes, mainly because the host is mobile and ensures an effective dispersal of its parasites. Nevertheless, some animal-parasitic nematodes show adaptations in their life cycles that facilitate the transfer from one definitive host to another. Earthworms are particularly important as transport hosts for certain parasites of birds (*Syngamus, Heterakis*). This particular relationship is not obligatory but the eggs ingested by the earthworm hatch and the juveniles can survive encysted in its tissues for long periods. Survival of the nematode is enhanced and the primary hosts increase the chance of infection by actively searching for earthworms during feeding.[232] Other nematodes have a more obligatory relationship and develop within a secondary host. Female *Dracunculus* behave in a remarkable manner, releasing juveniles from an open sore that they have produced when this part of the skin of the host is immersed in water. Further development of the juveniles released into water only occurs if they are ingested by *Cyclops*; mammals are subsequently infected by swallowing this copepod in drinking water.[232]

Adult filarial nematodes produce large numbers of microfilariae. Specific ectoparasitic vectors are infected by feeding on the peripheral blood or tissues of the infected, definitive host and transmit the parasite to other mammals after a period of further development. The vector of *Wuchereria* is a mosquito, *Loa loa* is transmitted by the mango fly (*Chrysops*) and *Onchocerca* is carried by blackflies (Simulidae). The microfilariae of *Wuchereria* show an interesting diurnal change in their distribution within the circulatory system of their final host that favours their ingestion by a suitable insect vector. A greater proportion of the total population occurs in the peripheral capillaries at that time of the day when the vector feeds on the blood of the final host.[232] The physiology of the host plays some part in this circadian rhythm but little is known of the role played by the physiology and behaviour of the nematode.

Trichinella spiralis has more than one definitive host and is able to reproduce in many carnivores and omnivores including man. The juveniles encyst in the muscles of the host and infect predators or scavengers that eat the infected muscles. The parasite never has a free-living existence but it is still very mobile because it is carried around by its host.[232]

9 Sense Organs and Behaviour

9.1 Sense Organs

9.1.1 Introduction

Sense organs that respond to touch and to chemical stimuli play an important role in the behaviour of nematodes, as most of these animals live in the dark in soil, in mud, or in the organs and tissues of other organisms. A very small number of nematodes have structures which may play a role in sensitivity to light. Unfortunately very little physiological work has been done on nematode sense organs and in most cases their function has had to be assumed from their location on the body of the nematode, or from their structure.

Sensilla are simple types of sense organs involving only a few neurons. In insects a sensillum is a sense organ that consists of a sensory neuron, part of the cuticle associated with the sense organ, and its supporting cells. Many sense organs in nematodes have a similar structure but are referred to as papillae. This can be misleading when the sensory structures referred to are long setae or bristles rather than small protuberances. It would be more accurate to call these sense organs sensilla, as in insects.[59]

9.1.2 Mechanoreceptors

Tactile sense organs are associated with various regions of the body wall of nematodes but most are concentrated around the mouth (figures 7.4 and 9.1) and, in the case of males, around the cloacal region (figure 6.2*B*).[4,14] The sensory dendrites of these tactile sensilla terminate in a modified cilium, which is embedded in the cuticle of the body wall (figure 9.2), is closely associated with the basal

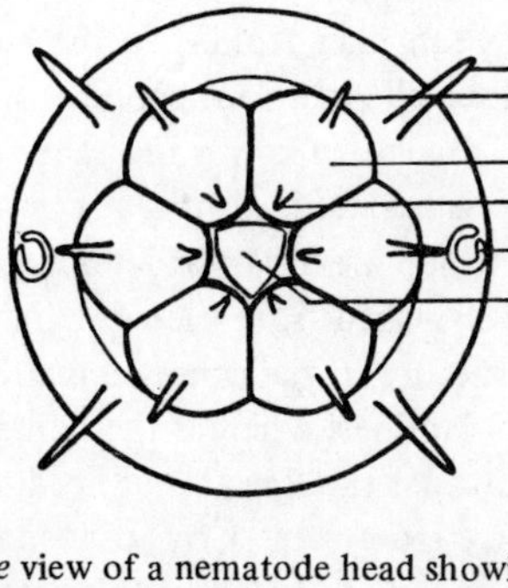

Figure 9.1 *En face* view of a nematode head showing the positions of the mouth, lips, amphids, labial sensillae and cephalic sensilla (usually called papillae). [After Jones, *Plant Nematology*, (H.M.S.O. 1959).]

layers of the cuticle, or is inserted into the base of one of the cuticular setae that are present on the surface of many aquatic nematodes (figures 9.3 and 9.6).[14,59,233] The neurons are surrounded by sheath cells (figures 9.2 and 9.5),which may regulate the ionic environment surrounding the sensory neurons and thus affect their sensitivity. Socket cells (figures 9.2 and 9.5) also surround the sensillum and appear to have a skeletal function.[59] In most nematodes the tactile sensilla end in small protruberences of cuticle (papillae) which may vary in size from species to species and also from one region of the body to another.

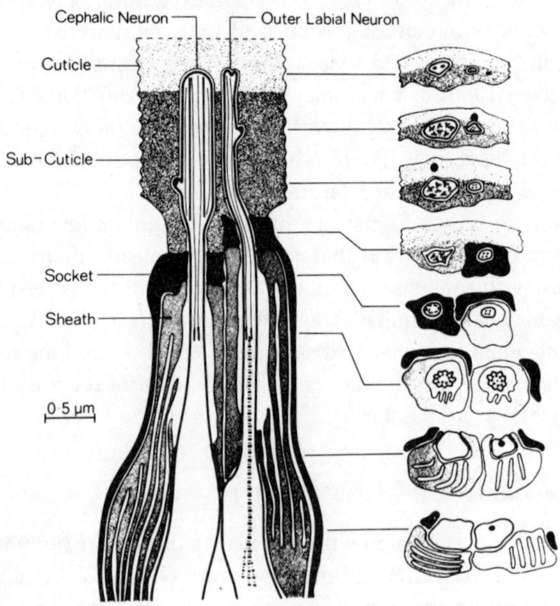

Figure 9.2 Nematode mechanoreceptors. Outer labial and cephalic sensilla of *Caenorhabditis elegans* shown in longitudinal and transverse section. (Ward, Thomson, White and Brenner.[59])

Most nematodes have three circles of sensilla (papilla) on the head (figure 9.1): six inner labial sensilla, six outer labial sensilla and four cephalic sensilla (figures 9.2 and 9.5). The ancestral pattern of labial sensilla was probably three to each of six lips but the lateral lips now possess two and not three sensilla (figure 9.1). It has been suggested that the deirids, which are a pair of lateral, cervical sensilla, are cephalic sensilla that have become displaced from their original position on the lateral lips as they are similar in structure to the other cephalic sensilla and both types of sensillum contain catecholamines, whereas the labial sensilla do not (section 7.2.4).[59,212] In some nematodes the lips are reduced in number (adult *Ascaris* and *Heterakis* have three (figure 3.1*B*)), or are lost as in *Syngamus trachea*, *Ancylostoma duodenale* (figure 3.2*G* and *H*) and *Bunostomum trigonocephalum* (figure 3.1*A*), but the full complement of sensilla are usually present.

Meloidogyne incognita has six inner labial sensilla and four cephalic sensilla but no outer labial sensilla as their dendrites have merged into the inner labial sensilla.[4,14] In some nematodes (*Nippostrongylus brasiliensis*) the distal end of the sensory neuron is bulbous or discoid and this probably increases the area of sensitivity. Some sensilla on the head of adult *Nippostrongylus* are similar in structure to chordatonal sensilla in insects as a strand of material extends across the fluid-filled middle layer of the inflated cephalic cuticle from the tip of the dendrite to insert on the cortical layer of the cuticle. Pressure applied to any region of the inflated cephalic cuticle will stretch the strand and activate the sensillum.[211] The various types of sensilla on the heads of nematodes will assist the nematode in exploring the environment during host-finding, feeding, hatching and mating. Presumably the mechanoreceptors are activated when the sensilla touch an object.

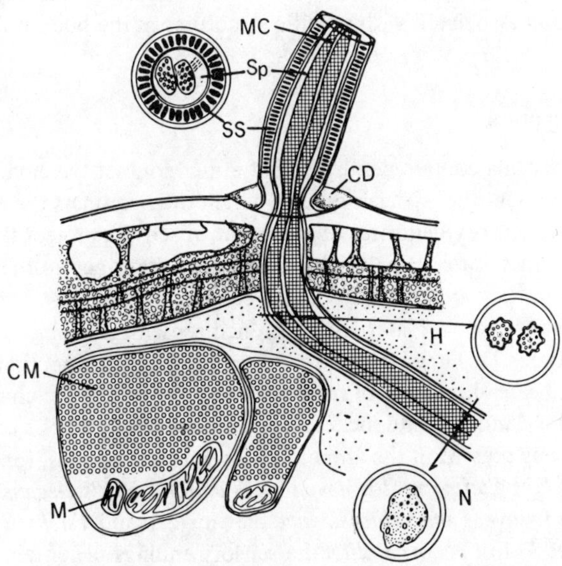

Figure 9.3 A somatic seta of *Chromadorina bioculata. CD* cuticular depression; *CM* contracting muscle; *H* hypodermis; *M* mitochondrion; *MC* modified cilium; *N* neuron; *Sp* space in seta; *SS* supporting strut (Croll and Smith.[233])

Deirids are tactile sensilla that are situated laterally on the body of the nematode in the region of the nerve ring. They occur on the sides of the body just before it reaches its maximum diameter (in most nematodes) and probably informs the nematode if it attempts to enter a pore or space that is too narrow for it to pass through. They are thought to be modified cephalic sensilla.[59]

Several genital sensilla are located around the cloacal region of males (figure 6.2*B*) and some are mechanoreceptors.[14] They consist of raised areas of thin cuticle that are innervated by one or more sensory dendrites. Some male nematodes possess a copulatory bursa at their posterior end (figures 1.1 and 6.3) and sensilla are situated at the ends of these bursal rays. The genital sensilla probably assist the nematode to orientate itself during copulation.

Most male nematodes possess one or two copulatory spicules in their cloaca. They dilate the vulva and vagina of the female prior to ejaculation of spermatozoa and in a few species (*Heterodera, Dipetalonema*) they also make a channel along which the spermatozoa can pass into the female. The spicules are innervated and the tips contain dendritic processes. In *Heterakis gallinarum* and *Nippostrongylus brasiliensis* the dendrites terminate just beneath the tip of each spicule but in *Heterodera rostochiensis* they project into two small pores near the tip of the spicule.[14,191] The sensilla in the tip probably enable the spicule to be inserted into the female without damaging her reproductive system.

Nematodes have many nerve endings beneath the cuticle and these seem to give sensitivity to the body wall. There are also a number of sense organs of unknown function associated with specific structures in the body wall of some nematodes.[14]

9.1.3 Chemoreceptors

Chemosensory sensilla contain dendrites that either contact the environment directly or contact another specialised cell, which then contacts the environment. In nematodes the sensory dendrites of a number of sensilla project through pores in the cuticle or open into cuticular pits or channels that open to the surface of the nematode and they are probably chemoreceptors (figures 9.4, 9.5).[14,59]

The amphids are a pair of laterally positioned sense organs on the head of nematodes (figure 9.1). They consist of a cuticular-lined channel that opens to the exterior on the head, a group of neurons that project into the channel, sheath cells (that may be glandular) and socket, supporting cells (figure 9.5). The number of sensory neurons present in the amphid varies with the species, for example, microfilariae of *Dirofilaria immitis* have nine, *Caenorhabditis elegans* juveniles and adults have twelve, *Panagrellus silusiae* has thirteen and *Trichodorus christei* has twenty-three.[14] In *Caenorhabditis* the sensory endings all contain basal bodies, nine outer doublet microtubules and two to six central single microtubules. Although most of the neuron endings in the amphid of this nematode are similar in morphology, each can be identified and occupies a particular position within the amphidial channel. Four of the amphidial neurons are unusually shaped and run along the amphidial channel for part of their length before terminating in invaginations of the sheath cell.[59] It is suspected that each sensory neuron in the amphid has a particular role to play in chemoreception. Amphids are apparently the main chemoreceptor sense organs on the anterior end of most nematodes but the only direct evidence that they are chemosensory is from experiments with *Caenorhabditis elegans*. Normal individuals respond to specific anions and cations, to basic pH and to cyclic AMP, but mutants that have no amphids do not respond to these stimuli (section 9.2.1)[59,234]

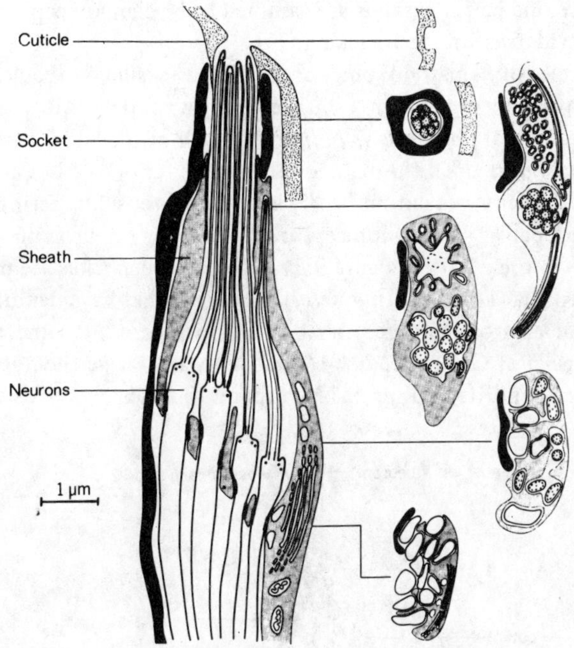

Figure 9.4 Nematode chemoreceptor. Diagram of an amphid, with its associated cells, from *Caenorhabditis elegans.* The amphid is shown in longitudinal and transverse section. The scale of the transverse sections is half that shown for the longitudinal section. (Ward, Thomson, White and Brenner.[59])

A gland is sometimes associated with the base of the amphidial channel. In *Caenorhabditis elegans* the sheath cell of the amphid contains a large Golgi apparatus and material from this is apparently secreted into the amphidial channel. Secretions from this Golgi apparatus may include mucus to protect the exposed dendrites.[59] The amphidial glands of other nematodes may also secrete a protective mucus but those of *Ancylostoma caninum* and of *Syngamus trachea* contain an anticoagulant, which is thought to be secreted by the nematode during feeding, and the glands of adult *Necator americanus,* which occupy about 25 per cent of the length of the worm, contain large amounts of acetylcholinesterase, which is passed along the amphidial channel to the exterior. The role of this enzyme in the biology of the nematode is not known.[14]

Phasmids are paired, lateral sense organs that are similar to the amphids in structure but are smaller and are situated behind the anus. Many free-living and a few plant-parasitic nematodes lack phasmids and this has been used as a diagnostic feature in the taxonomy of nematodes. In general the phasmidial channel contains fewer sensory dendrites than the amphids (*Dipetalonema witei* has one and *Necator americanus* has two) and the gland, when present, is much smaller. Some species in the genus *Hoplolaimus* have several pairs of phasmids but most nematodes have one pair. Phasmids are assumed to be chemoreceptors but physiological evidence for this is lacking.[14]

In some species of nematode some, or all, of the sensilla on the head open to the external environment through a small pore in the cuticle. All of the labial and cephalic sensilla of *Capillaria hepatica*[235] and all of the labial sensilla of adult *Meloidogyne incognita* open through pores to the exterior.[4,14] In *Caenorhabditis elegans* one of the sensory neurons of the inner labial sensillum terminates in the cuticle and is probably a mechanoreceptor, but the other neuron projects slightly through a pore to the exterior (figure 9.5).[59] Exposed dendrites are present in some genital sensilla (*Dipetalonema witei*).[14] Sensilla that have dendrites in direct contact with the environment are probably chemosensory but some, such as the inner labial sensilla of *Caenorhabditis elegans*, appear to have chemosensory and mechanosensory dendrites and probably respond to touch as well as to chemical stimulation.

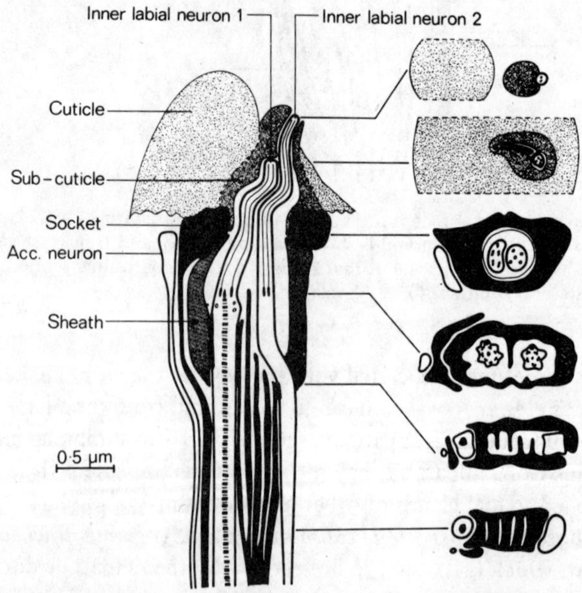

Figure 9.5 An inner labial sensillum from *Caenorhabditis elegans.* Note that inner labial neuron 1 ends within a pit in the cuticle and is probably a mechanoreceptor whereas inner labial neuron 2 is exposed at its tip through a pore in the cuticle and is probably a chemoreceptor. (Ward, Thomson, White and Brenner[59].)

The plant-parasitic nematode *Longidorus leptocephalus* appears to have gustatory sense organs associated with its buccal cavity and with its stylet. A group of sensory dendrites are associated with the anterior end of the buccal cavity and a second group is located within the three sinuses of the stylet where they are closely associated with pores that penetrate almost to the food canal. These sensory structures may have a gustatory function and may also facilitate the accurate location of feeding sites deep within the plant roots.[14]

9.1.4 Photoreceptors

Nematodes, like many other invertebrates, have a general sensitivity to light, but assumed photoreceptors have been found in very few species.[14,236]

Oncholaimus vesicarius has red granules of pigment within the anterior radial muscle cells of the pharynx. No nerves are associated with them but they lie close to the amphids and the pigment may act as a shade for light-sensitive receptors in the amphids. Haemoglobin is present in a structure called the chromotrope in the anterior end of gravid female *Mermis subnigrescens*. It is not present in the young female. This pigment apparently assists the nematode to regulate oviposition and also helps it to orientate itself when it ascends the aerial parts of plants. It is thought to function by shading certain areas of the nervous system.

More highly developed structures, which are presumed to be light sensitive, are found in certain other nematodes. Almost all nematodes that possess so-called ocelli are marine species (for example, *Acanthonchus rostratus, Deontostoma californicum*). Most of the ocelli are associated with the pharynx of the nematode and consist of a cup-shaped invagination containing some or all of the following: a pigment, a nerve, concentric lamellae which resemble the photoreceptive rhabdomeres found in some other invertebrates, and a lens which is usually formed from an invagination of the cuticle (figure 9.6).

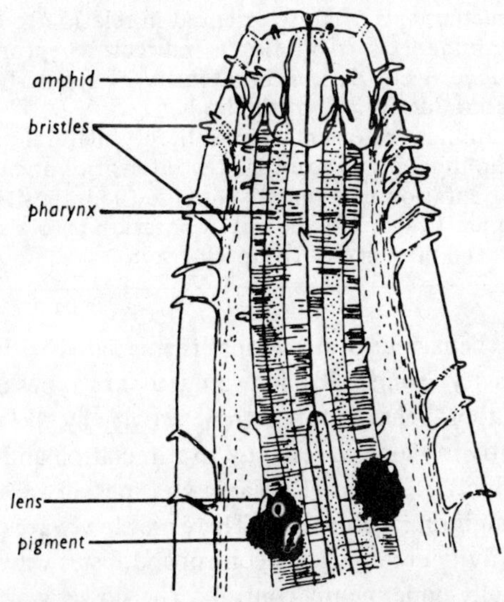

Figure 9.6 Head end of *Deontostoma californicum* (dorsal view) showing the position of the ocelli. Several sensory bristles associated with the anterior end of the nematode are also shown. [After Steiner and Albin, *J. Wash. Acad. Sci.* 23 (1933).]

The pigments are thought to aid directional sensitivity but photoresponses have not been detected in any nematode that possesses so-called ocelli. These ocelli have been called light-sensitive structures on anatomical and not on physiological evidence. It is interesting to note, however, that most nematodes that possess these structures live in the intertidal zone where the ability to detect changes in light intensity may be of use to the nematode.[236]

9.2 Behaviour

The behaviour of a nematode depends upon the internal state of the individual and a number of environmental factors. A single stimulus may initiate a response through a sense organ but some environmental changes may act directly on the nervous system or on the metabolism of the nematode. The simple behavioural response of a nematode may be undirected movement under particular stimulation (kinesis) or movement directed with respect to the source of stimulation (taxis, table 9.1).

Table 9.1 Simple Animal Movements (Prosser[164])

kinesis	undirected; no orientation of axis of body in relation to the stimulus
ortho-kinesis	speed or frequency of locomotion dependent on intensity
klino-kinesis	frequency or amount of turning per unit time dependent on intensity
taxes	directed; axis of body oriented in relation to the stimulus
klino-taxis	attainment of orientation is indirect; receptor is subjected to successive variations in intensity by regularly alternating lateral deviations of the body
tropo-taxis	attainment of orientation is direct; bilateral pair of receptors simultaneously compare the intensities on the two sides; orientation movements equalise the intensities
telo-taxis	similar to tropo-taxis, but orientation is to a single source even in the presence of competing sources

Much work on the behaviour of nematodes is associated with the analysis of their locomotory activity. Nematodes show a number of types of body wave that have distinct roles in their behaviour. Searching activity by the anterior region of the animal is frequently initiated in response to stimulation and plays an important part in feeding (section 3.2.2). There are two types of wave that pass backward along the animal during locomotion. There are slow waves, which are believed to be primarily under myogenic control and faster waves, which are thought to be essentially under neural control.[9] The slower wave form is frequently associated with movement over surfaces, and the faster form is recorded most frequently when nematodes swim. Body waves that travel anteriorly along the nematode also occur and are frequently the chief disruptive element in the symmetry of the locomotory path formed on a surface.[237] The sinusoidal track produced by the nematode usually makes an irregular pattern, but its form is modified in response to environmental stimuli. However, the over-all pattern produced by some species (*Caenorhabditis, Ancylostoma*) is characteristic of the individual even when responding to external stimuli. This suggests that these patterns are also influenced by endogenous factors such as the bias of body in undulations to one side in a way that results in differences between the patterns produced by nematodes of the same species.[237,238] Also, the individual may not always respond to a particular stimulus in a similar way on two separate occasions because its physiological condition may alter, particularly during development. Some well-developed responses to a single stimulus will be considered before an attempt is made to show how behavioural patterns arise from an integration of these reactions.

9.2.1 Responses to Chemicals

Chemoreception has a number of important features. The stimulus can vary in quantity (concentration) and quality (chemical structure) and so can provide a detailed basis for behavioural responses in animals equipped with adequate receptors; chemoreception is frequently an important form of sensory input in soil-inhabiting animals. The available evidence suggests that feeding, reproduction and host location in some nematodes are partly dependent on chemoreception. In feeding, many of the stimuli may be chemicals with some nutritional status but in other cases the chemical is a token stimulus that is of no direct value to the animal. The receptors initiating a response may be sensitive and respond to chemicals diffusing from a distant source (olefaction) or they may be less sensitive and respond to contact with the source (taste). Escape responses may be initiated by relatively insensitive and non-discriminating chemoreceptors.[164] Work with males of *Heterodera* has shown that this nematode orientates to virgin females of this genus. Furthermore, this would seem to be a response mediated by chemoreceptors of an olefactory nature that respond to a token stimulus. Since this suggests that the most sensitive form of chemoreceptor is present in nematodes it is probable that they also possess the two less sensitive types.

The response of male *Heterodera* to females is also important as it forms one useful hypothesis for orientation of a nematode to a stimulus. Virgin female *Heterodera* release chemicals (section 6.1.2) and on agar plates approximately radial diffusion gradients are produced by centrally placed females. The behaviour of the males in the presence of these females has been studied in detail and has also been filmed. Each male moves by sinusoidal waves but apparently with a bias that results in a broad arc rather than a straight path. Periodically the nematode will tend to move down the concentration gradient. This causes the nematode to stop; it then appears to sample around itself with the anterior end and moves off in any direction which has a greater concentration of the chemical. In this way the animal forms a series of irregular arcs and tends to produce a spiral path towards the female (figure 9.7). This seems to be a negative klinokinesis, with increased turning produced by a reduction in chemical concentration.[16,193] Clearly, movement in soils is more complex than that on agar plates, for location of the female must take place in three dimensions. The tortuous nature of pore channels in soils might ensure that chemicals diffuse readily from a source down channels narrower than those available to the nematode. The klinokinetic response may be particularly suited to these conditions. The nematode may move down any channel suitable for movement but this simple co-ordination would eventually lead the animal through the labyrinth to the female. It is unknown if this proposed klinokinesis occurs frequently in soil nematodes and it certainly cannot be assumed that it is a typical response to chemicals for nematodes from other habitats.

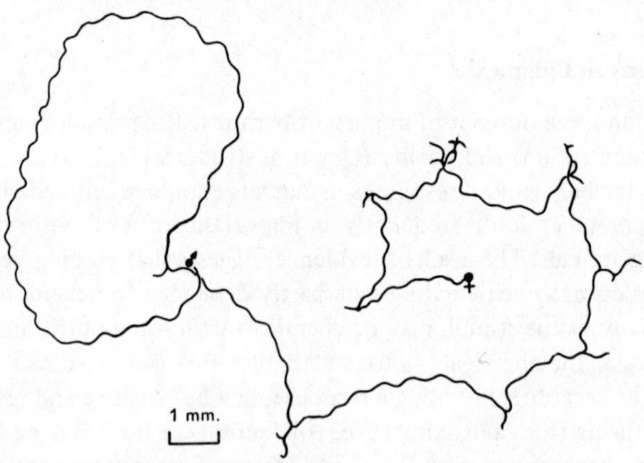

Figure 9.7 *Heterodera schachtii* male approaching a female by a spiral, polygonal track after a preliminary circle. The male was placed at ♂ and the female at ♀. (Green.[193])

Work on agar plates has suggested that a different mechanism is involved in the bacterial feeder *Caenorhabditis*. This nematode is attracted to cyclic nucleotides, certain anions and cations and also to basic pH[234] (table 9.2). Further work using a different experimental approach has shown that *Caenorhabditis* also avoids acid pH and the response to hydrated forms of carbon dioxide at concentrations normally found in soils is dependant on the buffer used.[239] The response to cyclic AMP is of particular interest because the chemical is apparently released by many bacteria and could be a fairly specific indication of the location of food. The amphids are involved in this response and forms lacking functional amphids fail to respond to the attractants. However, the response is a taxis and not a kinesis. Mutants with heads bent to one side apparently maintain the head in the direction of the gradient, indicating a tactic response of the head region to

Table 9.2 Summary of attractants for *Caenorhabditis* (Ward[234])

Class	Attractant	Accumulation threshold (mM)
1. Cyclic Nucleotides	cAMP	0.2
	cGMP	0.2
	Not AMP, Bu_2–cAMP	
2. Anions	Cl^-	2
	Br^-	20
	I^-	20
	Not CH_3COO^-, F^-	
3. Cations	Na^+	2
	Li^+	4
	K^+	15
	Mg^{2+}	20
	Not $NH_4{}^+$, $CH_3NH_3{}^+$	
4. Basic pH	OH^-	~ 0.001
Unclassified	lysine	~10
	histidine	~10
	cysteine	~10

The attractants are shown grouped into classes, according to competition experiments. The accumulation threshold is the concentration of attractant at the centre of the plate that causes 25 per cent of the population (twice the background) to accumulate. Because of their high threshold, the amino acids have not been extensively studied except to show that the amino acids, and not their counter-ions, are the attractants.

the concentration gradient. During locomotion these malformed individuals make precise and continual compensation for a slewing effect caused by the direction of motion of the body lying at an angle to the bent head. The response is believed to be a klinotaxis with the amphids acting together during the swing of the head. Tropotaxis is not indicated for a few reasons. Nematodes move by dorso-ventral undulations. The amphids, therefore, lie perpendicular to the plane of motion and consequently are unsuited to function as the paired sense receptors required for tropotaxis (table 9.1). A second receptor might be located in the tail but this seems unlikely because mutants lacking functional phasmids orientate normally.[234]

The mechanism of orientation in *Caenorhabditis* is apparently different from the response that ensures that they remain at a located source of attractant. This is an important point, for it may be that arrival at the food or host initiates a further stage in a behavioural sequence and this further stage may be dependent on different stimuli. In feeding, contact chemoreceptors and tactile stimuli may prove to be important after the nematode has located its food[234] (section 3.2.2).

The chemokinetic or chemotactic response of other nematodes is less well understood. They undoubtedly do occur and sex attractants are believed to be produced by a range of nematodes (*Panagrolaimus, Pelodera, Heterodera, Nippostrongylus*).[240,16,241] Furthermore, plant roots are considered to attract some plant-parasitic nematodes. Specific chemicals may be important in some cases but carbon dioxide seems to be an important stimulus for many species.

9.2.2 Response to Mechanical Stimulation

Most nematodes show a sensitivity to mechanical stimulation. Some species are thigmokinetic (*Enoplus communis, Nippostrongylus*) and a number of individuals of either species placed in saline will remain hyperactive until a medusoid mass is formed.[246] Such a response may be important in the normal habitats of these animals. *Enoplus communis* must withstand wave action of the sea in its habitat among holdfasts of algae and between *Mytilus* shells on a rocky shore. Its tendency to entwine itself with other objects has an obvious value in maintaining its position in this habitat. The related *Enoplus brevis* does not show a marked thigmokinetic response and this can be correlated with its habitat in marine mud that is rarely disturbed by the sea. The enhanced activity of *Nippostrongylus* adults following their loss of contact with the mucosa favours the possibility that they will regain their normal position before they are swept along the lumen of the gut away from the intestine.[245,246] Individuals of many species tend to aggregate as the result of thigmotactic or thigmokinetic behaviour. This may ensure that a displacement of some species involves a reproductive unit of males and females rather than just an individual.

Many free-living nematodes become hyperactive if stimulated by touch. *Rhabditis* sp. also shows a further response: it moves backwards if touched near the anterior end and in the opposite direction if stimulated near the tail region.[240] These reactions to touch represent a simple escape response that may be related to predation of this species in soil.

9.2.3 Reactions to Heat

Environmental temperature limits the distribution of animals and also affects their activity. Nematodes are poikilotherms, consequently changes in temperature will influence the rate of their biochemical processes. Some control of these effects is possible through metabolic compensation, through dormancy reactions (particularly in response to low temperatures) and by behavioural responses. Unfortunately it is not always possible to determine if changes in activity of a nematode are due to a direct effect of heat on metabolism, on the excitability of nerves, or if the reaction is mediated through specialised, heat-sensitive receptors.

It would seem that many of the reactions of nematodes to heat are due to the direct effects of heat on animal cells. Some nematodes may possess thermoreceptors but these sense organs have a simple structure in many animals and are not readily identified. A number of nematodes do react to temperature gradients. Adult *Nippostrongylus* accumulate at the hotter end of temperature gradients ranging from 28–48 °C or from 37–52 °C, although the latter upper limit can result in death of the individual due to thermal damage. Possibly this reaction is not a natural effect because marked thermal gradients and lethal temperatures are not found in the intestinal mucosa of the rat.

Not all nematodes respond to temperature in this way. Some show an eccritic (preferred) temperature within their normal environmental range. Fourth-stage juveniles of *Phocanema* (*Terranova*) *decipiens* accumulate at 32.5 °C when placed in either a hotter or cooler position in a thermal gradient. They are most active and survive longer at this temperature. However, the avoidance of certain temperatures in favour of others does not necessarily involve complex temperature sensors, as *Paramecium* also shows an eccritic temperature response.[240]

Ditylenchus dipsaci tends to favour temperatures below 25 °C in a thermal gradient but its distribution may be influenced by previously experienced temperatures. Field populations from upland soils in parts of America invade lucerne at lower temperatures than individuals recovered from warmer, lowland regions. However such reactions may not be entirely due to heat because temperature also influences the plant and this could affect its suitability for nematode populations of different origins.[242]

Infective juveniles of nematodes that penetrate the skin of homeotherms (*Nippostrongylus, Ancylostoma, Necator*) show a marked response to temperature. Activity of these species is increased enormously by proximity to objects with a temperature similar to that of their hosts. This clearly has an adaptative significance in host location and is also of interest because temperature is acting in a manner similar to a token stimulus. In *Nippostrongylus* an increase in ambient temperature to approximately that of the host is initially accompanied by a large increase in oxygen consumption. This indicates an exceptionally high and transient effect of temperature on the level of metabolism, which is probably orthokinetic in nature and it need not, therefore, involve sensory receptors.

9.2.4 Responses to Light

Light is both a quantitative and qualitative stimulus to animals. However, the responses of nematodes to this stimulus are considerably more limited than those of many more advanced invertebrates.

There are a number of recorded responses of nematodes to light. A species of *Rhabditis* swarms on the illuminated surface of mushroom beds but disappears from the surface in the dark.[240] Similarly, juveniles of *Dictyocaulus viviparus* apparently move towards light and climb sporangia of *Pilobolus* prior to their dispersal by this fungus[230] (section 8.4).

Much early work on the responses of nematodes to light has been severely questioned following subsequent demonstrations that heat radiating from the light source was responsible for many of the recorded responses.[240] Nevertheless, many invertebrates do show photokinetic responses without clearly definable multicellular receptors. The exact nature of these responses is unknown but in some cases the receptor is considered to be within the central nervous system. Simple photoreceptors may be surrounded by a cup-shaped or other asymmetrical layer of pigment. This arrangement can even be the basis for a tactic response, for the receptor will receive a reduced light intensity whenever the pigment lies between it and the light source. A lens provides greater light-gathering ability, but more advanced receptors differ in other ways. In particular, the associated pigment is photolabile and the light initiates a series of chemical reactions that trigger the receptor cell.[164]

Some marine nematodes do contain pigments which are apparently associated with ocelli (section 9.1.4). The animals have not been shown to be phototactic but the ocelli are particularly correlated with forms that live in habitats exposed to light. Ultrastructural studies suggest that some are simple ocelli and they probably lack a photolabile pigment.[236] One approach to demonstrating photoreception is to construct an action spectrum of a possible photoreceptor which relates the number of photons required for a constant response at each wavelength. This action spectrum is then related to the probable chemical nature of the pigment associated with the photoreceptors.[164] There has been no comprehensive study of any nematode in this way but some preliminary experiments have been carried out on *Chromodarina* and *Mermis*. Apparently, the only discrete pigment in both nematodes is a haemoglobin. *Mermis* ascends herbage to lay eggs and the uterine contractions which accompany oviposition are particularly stimulated when the animal is exposed to green light. The response could have adaptative significance as light shining through leaves may serve as a stimulus for egg laying. It has been suggested that this response is correlated with the greater absorption of light by haemoglobin from the green portion of the spectrum.[240] If haemoglobin does function in this way it seems likely that it shades the receptor causing stimulation. Haemoglobin is not a recognised photochemical *in vivo* and it seems unlikely that it has a photolabile function.

9.2.5 Responses to an Electrical Field

Many free-living and plant-parasitic nematodes have been reported to show movement towards the cathode (*Panagrellus, Turbatrix, Rhabditis, Dorylaimus, Tylenchus, Ditylenchus, Aphelenchoides*) or the anode (*Heterodera schachtii* juveniles). There are difficulties associated with these findings. Electrical fields are not usually considered a normal stimulus, mainly because appreciable gradients and current intensities do not usually act on animals in their natural environments. It has been suggested that plant-parasitic nematodes are exceptions to this generalisation. Plant roots maintain electric potentials across their epidermis of up to approximately 60 mV.[198] Much of the potential difference occurs across the cell walls and may be associated with the uptake of minerals by the plant. It is possible that the animals do respond to these gradients when in contact with the plants. This potential frequently alters along the root. Possibly the animals can sense these changes and utilise the stimulus in locating a preferred zone for root invasion or feeding.[16] Some plant-parasitic nematodes also respond to oxidation–reduction potentials, but this response may be to changes in the available ions associated with alterations in this potential.

9.2.6 Responses to Gravity

Animals may possess sense organs that respond to acceleration of the animal or static receptors that convey information about gravity.[164] There is no evidence for the dynamic type of receptor in nematodes and it is far from certain that they respond to gravity. Gravity is virtually constant in magnitude and direction, therefore animals may respond to this stimulus with a taxis but, by definition, they cannot show a kinetic response. Free-living stages of many animal-parasitic nematodes often ascend herbage, but upward movement of the juveniles is the result of random movements and not a response to gravity. The juveniles move in all directions, even on a vertical plane. The moisture on the grass in their habitat and the nature of the leaf surface are important modifiers of their upward or downward movement[222] (section 8.2.2). Similarly, there is no evidence at present that the upward movement of nematodes that are parasitic on the aerial parts of plants is associated with a negative geotaxis. Finally, nematodes feeding at the surface of deep fluids must overcome a tendency to sink. *Turbatrix aceti* inhabits the surface of malt vinegar but does not orientate to gravity. It maintains its position at the surface because it is an active swimmer with a centre of gravity within the posterior half of its body and it does not, therefore, readily sink in fluids.

9.2.7 Response to a Moisture Gradient

Active nematodes in the soil are always in water and cannot be hydrotactic. Juveniles of *Heterodera rostochiensis* are believed to be attracted to roots along a moisture gradient but this could be a mechanical response or the animal may be responding to chemical gradients in the rhizosphere of the plant.

9.3 Whole Patterns of Behaviour

In natural conditions nematodes are likely to experience a number of environmental stimuli, some of which vary with time. Furthermore, the animal shows internal changes during its development and these may modify its receptivity to particular changes in its environment. A few better-studied examples can show how the responses to specific stimuli are integrated into the whole patterns of behaviour shown by an individual in its niche.

9.3.1 Location and Penetration of Host Plants

The movement of plant-parasitic nematodes to their host frequently involves an appreciable random component. Movement of *Aphelenchoides fragiarae* up plants to leaves is mainly dependent on suitable environmental conditions (section 8.2.2). The animal moves randomly on the leaf until it reaches an interruption in the water film around a stoma. The animal responds to mechanical stimulation of its head by the film surface and starts to probe with the head region. Entry into the leaf through the stoma is a response to the diffusion of carbon dioxide from these pores, as the nematode will even enter micropores in agar discs providing they are emitting this gas.[243]

A random component to host finding is involved in most root parasites but chemoreceptors are probably important when the nematode is close to the roots. Gradients of carbon dioxide concentrations may often be a main attractant to nematodes but some species fail to show a marked aggregation around roots of non-host plants and may be influenced by specific stimuli present near the roots of their hosts (*Heterodera schachtii, Meloidogyne hapla*).[240]

A few nematodes may respond to roots over considerable distances. *Trichodorus* has been studied around apple roots in an orchard by time-lapse photography through glass-sided trenches. The growing root tips show an apparently changing attractiveness to the nematodes (figure 9.8). The population is mainly

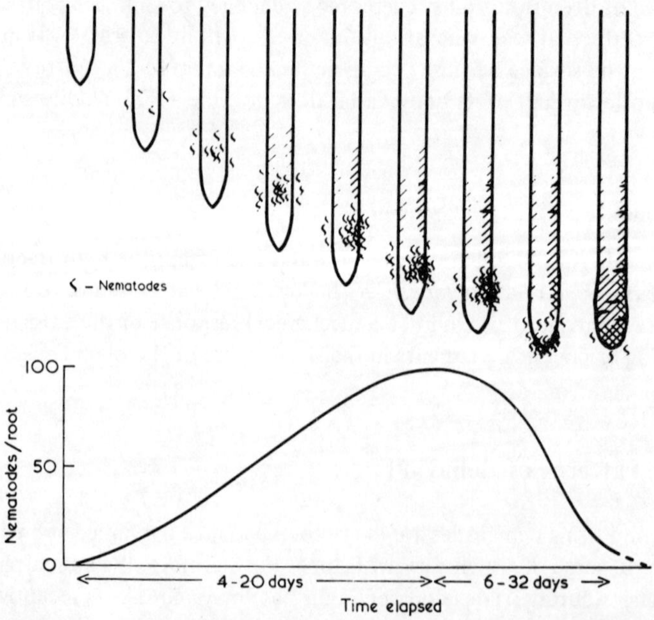

Figure 9.8 The accumulation of *Trichondorus viruliferus* around the extending roots of the apple. The shaded area indicates the extent of the necrosis caused by the nematodes. (Pitcher.[244])

of adults and the increase in population is too rapid to be explained by multiplication at the root. It seems that most of the individuals in the soil within at least an 8 cm radius must aggregate at the root-tip during the period of attraction. Possibly the rate of accumulation is not orthokinetic in nature but results from either a chemokinetic or chemotactic response to roots.[244]

The behaviour of the plant-parasitic nematode changes when it reaches the root. Juveniles of *Heterodera* show a widespread exploration of a potential invasion site, followed by more localised head movements culminating in stylet thrusting (section 3.2.2). There is no direct experimental evidence but it seems that the animal can assess the nature of the surface. Tactile receptors, probably the labial sensilla, may provide information on the contours of the root surface. Deformation of the surface, which follows eversion of the stylet, may be detected by some form of proprio-receptor sited in the somatic or the stylet protractor musculature. Contact chemoreceptors may also be involved at this stage of the process. There is no doubt that *Heterodera* shows a high degree of coordination during penetration of plant cells. There are also marked similarities between this over-all behaviour and that shown by the juvenile during hatching (section 6.2.1), and by the male as it emerges from within the cuticle of a previous juvenile stage.[66]

9.3.2 Location and Penetration of Host Animals

Infective juveniles of some animal-parasitic nematodes penetrate the skin of their host. Many (*Nippostrongylus, Necator, Ancylostoma*) remain relatively inactive until heat from a nearby homeotherm triggers increased movement. The activity of the infective juveniles of *Ancylostoma* and *Trichonema* is minimal in constant conditions but a period of movement is stimulated by even a transient mechanical stimulation or by an increase in light levels. These responses may have adaptative significance because the juveniles will tend to conserve energy reserves by suppressing high levels of motility except in response to stimuli that may indicate the proximity of a potential host.[238]

Infective juveniles of *Nippostrongylus* on a moist surface tend to raise the anterior part of their head and perform a sinusoidal, searching activity. The combination of these responses enhances its chances of becoming attached to the hairs of a rat. If this occurs, the animal responds to the tactile stimulus of the hair by spiralling along it to the skin of the rat. The animal moves onto the skin and does not penetrate hair follicles; it moves over the surface in a random manner and eventually enters a crevice, perhaps in response to tactile stimulation. The greater purchase available in a crevice enables the juvenile to push the stratified keratin apart and allows entry into the skin[77] (section 8.2.4). Little is known of the behavioural responses of *Nippostrongylus* in the host during migration to the niche exploited by adults. However, the adult shows a marked behaviour that enables it to maintain its position in the niche. If displaced along the intestine the adults apparently move back to their original position. They are also thigmotactic or thigmokinetic and remain close to the mucosa of the host.[245,246]

These few examples show the nature of the responses of nematodes to changes in their environment. The animals respond in a basically simple manner, but the significant environmental stimuli differ from one habitat to another. Combinations of stimuli and specificity, particularly to chemicals, are probably the basis of the unique behavioural patterns favouring the survival of each species in its niche.

Appendix

A list of nematodes that are mentioned in this book with some short notes on their biology.

Nematode	Biology
Acanthocheilonema perstans	Adult parasitic in man (filarial nematode); juveniles (microfilariae) transmitted by *Culicoides* (midges).
Acanthonchus duplicatus	Free-living; lives around holdfast of sea-weed.
A. rostratus	Similar to *A. duplicatus.*
Acrobeles spp.	Free-living in soil; microbivorous.
Actinolaimus spp.	Free-living; predatory.
Agamermis spp.	Adult free-living; juvenile parasitic in the haemocoele of insects.
Ancylostoma caninum	Adult parasitic in intestine of dog (hook-worm); first three juvenile stages free-living; third juvenile penetrates skin of host.
A. duodenale	Similar to *A. caninum* but parasitic in intestine of man (hookworm); feeds on mucosa.
Angiostrongylus cantonensis	Lungworm of rat; infective juveniles in crustacea.
Anguina tritici	Parasitic on plants (cause of ear cockle in wheat).
Angusticaecum spp.	Parasitic in alimentary canal of tortoise.
Anisakis spp.	Parasitic in stomach of marine mammals and birds.
Aphelenchoides spp.	Some species are plant parasites; some feed on fungi; some are parasitic in insects.
Aphelenchoides blastophorus	Mycophagous

Nematode	Biology
A. composticola	Mycophagous; pest of mushrooms.
A. fragiarae	Plant parasitic.
A. parienturis	Mycophagous (on lichen).
A. ritzemabosi	Plant parasitic.
Aphelenchus avenae	Mycophagous.
Aplectana spp.	Parasitic in Amphibia.
Aporcellaimus spp.	Carnivorous.
Araeolaimus elegans	Marine mud.
Ascaridia galli	Parasitic in intestine of fowl.
Ascaris columnaris	Parasitic in intestine of skunk.
A. lumbricoides	Parasitic in intestine of man (and pig?).
A. suum	Parasitic in intestine of pig.
Aspiculuris tetraptera	Parasitic in colon of rodents.
Bradynema spp.	Adult parasitic in haemocoele of insects.
Bunostomum trigonocephalum	Parasitic in intestine of sheep and goats (hookworm).
Brugia malayi	Lymphatics of man and other mammals (filarial nematode); juveniles (microfilariae) circulate in blood; transmitted by mosquito.
B. pahangi	Similar to *B. malayi* but found mainly in carnivores.
Caenorhabditis briggsae	Free-living in soil; microbivorous.
C. elegans	Similar to *C. briggsae.*
Camallanus spp.	Parasitic in alimentary tract of fish, amphibia and reptiles.
Capillaria hepatica	Parasitic in liver of mammals.
Chabertia spp.	Parasitic in alimentary tract of ungulates.
Chaetosoma spp.	Free-living; marine.

Nematode	Biology
Chromadorina bioculata	Free-living in fresh water.
C. germanica	Free-living in fresh water.
Cooperia punctata	Parasitic in intestine of ungulates.
Cosmocerca ornata	Parasitic in intestine of amphibia.
Criconemoides spp.	Ectoparasitic on plant roots (ring nematodes).
Cyathostoma lari	Head sinuses of birds (herring gull).
Cyathostomum spp. (see *Trichonema*)	Parasitic in caecum and colon of Equidae.
Deladanus siridicola	Free-living generation mycophagous; parasitic generations in haemocoele of *Sirex* (wood wasp).
Deontostoma californicum	Free-living; marine.
Desmoscolex spp.	Free-living in fresh water.
Dictyocaulus viviparus	Parasitic in lungs of cattle.
Dioctophyme renale	Parasitic in kidney of mammals.
Dipetalonema witei	Filarial nematode.
Diplogaster spp.	Some free-living (microbivorous or predators); others parasitic in insects.
Dirofilaria immitis	Parasitic in heart and connective tissue of mammals; juvenile (microfilariae) circulate in blood; transmitted by mosquito.
D. insignia	Filarial nematode.
D. uniformis	Filarial nematode.
Ditylenchus destructor	Plant parasite (potato rot nematode).
D. dipsaci	Plant parasite (bulb and stem nematode).
D. myceliophagous	Mycophagous.
D. reformis	Mycophagous.
Dorylaimus spp.	Predatory and phytophagous species occur in soil and fresh water.

Nematode	Biology
Dracunculus medinensis	Parasitic in connective tissue of mammals (Guinea worm); juvenile in copepods.
Enoplus brevis	Free-living; marine.
E. communis	Free-living; marine.
Eudorylaimus spp.	Free-living in wide range of habitats; feeds on unicellular algae, protozoa, fungi, mite eggs.
E. obtisicaudatus	See *Eudorylaimus* spp.
Enterobius vermiculuris	Parasitic in colon of man.
Eustrongylides ignotus	Parasitic in intestine of birds; juvenile in muscles of fish.
Graphidium strigosum	Parasitic in alimentary tract of rabbit.
Haemonchus contortus	Parasitic in abomasum of sheep and cattle (stomach worm); first three juvenile stages free-living.
Hammerschmidtiella diesingi	Parasitic in hind-gut of cockroaches.
Hemicycliophora arenaria	Ectoparasitic on plant roots (sheath nematode).
Heterakis gallinarum	Parasitic in caecum of fowl.
Heterodera avenae	Endoparasitic on roots of cereals.
H. cruciferae	Endoparasitic on roots of cruciferae.
H. göttingiana	Endoparasitic on roots of peas (pea cyst nematode).
H. rostochiensis	Endoparasitic on roots of potatoes and tomatoes, (potato cyst nematode or golden nematode).
H. schachtii	Endoparasitic on roots of sugar beet and other hosts (sugar beet cyst nematode).
Hirschmaniella gracilis	Plant parasitic.
Hoplolaimus spp.	Plant parasitic; attack roots.

Leidynema appendiculata	Parasitic in hind-gut of cockroaches.
Litomosoides carinii	Parasitic in body cavity of rodents; transmitted by ticks.
Longidorus spp.	Plant-parasitic; attack roots; some species are vectors of viruses.
Meloidogyne hapla	Endoparasitic in plant roots (root-knot nematode).
M. incognita	Similar to *M. hapla*.
M. javanica	Similar to *M. hapla.*
Mermis nigrescens	Adult free-living; juvenile parasitic in haemocoele of insects.
M. subnigrescens	Similar to *M. nigrescens.*
Monhystera spp.	Free-living; phytophagous.
Mononchus spp.	Free-living; predator; cosmopolitan.
Mesodiplogaster lheritieri	Microbivorous; may be predator.
Naccobbus spp.	Endoparasitic on roots.
Necator americanus	Parasitic in intestine of man (hookworm); life cycle similar to that of *Ancylostoma*.
Nematodirus spp.	Parasitic in intestine of ruminants.
N. battus	Parasitic in intestine of sheep; egg requires exposure to cold before it will hatch.
Nematospiroides dubius	Parasitic in intestine of mice.
Neoaplectana carpocopsae	Parasitic in insects and feeds on their decaying tissues after killing insect.
Nippostrongylus brasiliensis	Parasitic in intestine of rats; life cycle similar to that of *Ancylostoma.*
Nygolaimus spp.	Free-living; predator.
Obeliscoides cuniculi	Parasitic in alimentary tract of rabbit.
Oesophagostomum radiatum	Parasitic in intestine of cattle.

Nematode	Biology
O. venulosum	Parasitic in colon of sheep, goat, deer.
Onchocerca volvulus	Parasitic in connective tissue of mammals; transmitted by *Simulium* spp.
Oncholaimus vesicarius	Free-living; marine.
Ostertagia spp.	Parasitic in abomasum of ruminants.
Oswaldocruzia spp.	Parasitic in alimentary tract of amphibia and reptiles.
Oxyuris equi	Parasitic in large intestine of horses.
Panagrellus redivivus	Free-living in sour paste; microbivorous.
P. silusiae	Similar to *P. redivivus.*
Panagrolaimus rigidus	Free-living; cosmopolitan.
Parascaris equorum	Parasitic in small intestine of horses.
Parasitylenchulus diplogenus	Parasitic in fruit flies (*Drosophila*).
Parasymphlocostoma formosum	Free-living; marine.
Paratylenchus nanus	Ectoparasitic on roots.
Passalurus spp.	Parasitic in alimentary tract of rabbit.
Pelodera chitwoodi	Free-living; microbivorous.
Phocanema decipiens	Parasitic in stomach of seals; juvenile in muscles of fish.
Pontonema vulgaris	Free-living; marine.
Porrocaecum decipiens (= *Phocanema*)	Parasitic in alimentary bract of seals; infective juvenile in fish.
Pratylenchus penetrans	Endoparasitic in plant roots (root lesion nematodes).
Radopholus similis	Endoparasitic in plant roots (burrowing nematode).
Rhabdias bufonis	Parasitic in lungs of amphibia.
Rhabditis coarctata	Free-living in dung; dauer juveniles transported by insects.

Nematodes	Biology
R. dubia	Free-living in dung; dauer juveniles transported by insects.
R. marina	Marine.
R. maupasi (= *R. pellio*)	Juveniles parasitic in earthworms; adults and juvenile stages live on decaying earthworm.
R. terrestris	Similar to *R. maupasi.*
Rotylenchulus spp.	Parasitic on plant roots.
Rotylenchus spp.	Parasitic on plant roots.
Seinura tennuicaudata	Free-living; predator.
Sphaerularia bombi	Parasitic in haemocoele of bees.
Spironoura spp.	Parasitic in intestine of fish, amphibia and reptiles.
Strongyloides fülleborni	Parasitic generations in alimentary tract of primates; free-living generations in faeces–soil mixture.
S. ratti	Similar to *S. fülleborni* but parasitic generations in rat.
Stephanurus dentatus	Parasitic in kidney of pigs.
Strongylus edentatus	Parasitic in large intestine of horses.
Syngamus trachea	Parasitic in bronchi of birds.
Syphacia obvelata	Parasitic in alimentary tract of rodents.
Tetrameres americana	Parasitic in proventriculus of poultry.
Thelastoma spp.	Parasitic in hind-gut of cockroaches.
Tobrillus aberrans	Free-living; microbivorous.
Toxocara canis	Parasitic in intestine of dogs.
T. mystax (= *T. cati*)	Parasitic in intestine of cats.
Trichinella spiralis	Parasitic in intestine of mammals; juveniles encyst in muscles of mammals.

Nematodes	Biology
T. vituliferus	Similar to *T. christei*
Trichodorus christei	Parasitic on plant roots; some species are virus vectors.
Trichonema (*Cyathostomum*) spp.	Parasitic in caecum and colon of Equidae.
Trichostrongylus axei	Parasitic in abomasum of sheep and cattle; first three juvenile stages free-living.
T. colubriformis	Parasitic in small intestine of sheep and cattle; first three juvenile stages free-living.
T. retortaeformis	Parasitic in small intestine of rabbit; first three juveniles stages free-living.
Trichuris muris	Parasitic in intestine of rats and mice.
T. suis	Parasitic in caecum of pigs.
T. vulpis	Parasitic in intestine of dog.
Turbatrix aceti	Free-living in malt vinegar; microbivorous (vinegar eelworm).
Tylenchorhynchus dubius	Ectoparasitic on plant roots.
Tylenchulus semipenetrans	Parasitic on plant roots.
Tylenchus spp.	Free-living or plant parasites; feed on root hairs or fungi.
Uncinaria stenocephala	Parasitic in intestine of dog (hookworm); third juvenile penetrates the skin.
Wuchereria bancrofti	Parasitic in connective tissue of man (filarial worm); juveniles (microfilariae) circulate in blood; transmitted by mosquitoes.
Xiphinema index	Parasitic on plant roots (dagger nematode); some species are virus vectors.

References

1 N. A. Cobb, 'Nematodes and their Relationships', in *U.S. Department of Agriculture Yearbook for 1914*, (1965) pp. 457–90.

2 J. E. Harris and H. D. Crofton, 'Structure and Function in the Nematodes: Internal Pressure and Cuticular Structure in Ascaris', *J. Exp. Biol.*, 34 (1957) pp. 116–30.

3 I. Zmoray and A. Guttekova, 'Ecological Conditions for Occurrence of Cilia in Intestines of Nematodes', *Biologia, Bratisl.*, 24 (1969) pp. 97–112.

4 A. F. Bird, *The Structure of Nematodes* (Academic Press, New York, 1971).

5 D. L. Lee, 'The Structure and Composition of the Helminth Cuticle', *Adv. Parasitol.*, 4 (1966) pp. 187–254.

6 D. L. Lee, 'The Structure of the Helminth Cuticle', *Adv. Parasitol.*, 10 (1972) pp. 347–79.

7 J. Rosenbluth, 'Ultrastructural Organization of Obliquely Striated Muscle Fibres in *Ascaris lumbricoides*', *J. Cell Biol.*, 25 (1965) pp. 495–515.

8 T. von Brand, *Biochemistry of Parasites*, 2nd ed. (Academic Press, New York, 1973).

9 H. D. Crofton, 'Form, Function and Behaviour', in *Plant Parasitic Nematodes.* I, eds B. M. Zuckerman, W. F. Mai and R. A. Rhode (Academic Press, New York, 1971) pp. 83–113.

10 B. G. Chitwood and M. B. Chitwood, *An Introduction to Nematology,* 2nd ed. (Monumental Printing Co., Baltimore, 1950).

11 D. L. Lee, 'The Fine Structure of the Excretory System in Adult *Nippostrongylus brasiliensis* (Nematoda) and a Suggested Function for the "Excretory Glands" ', *Tissue and Cell,* 2 (1970) pp. 225–31.

12 B. M. Ogilvie, T. L. W. Rothwell, K. C. Bremner, H. J. Schnitzerling, J. Nolan and R. K. Keith, 'Acetylcholinesterase Secretion by Parasitic Nematodes. I. Evidence for Secretion of the Enzyme by a Number of Species', *Int. J. Parasitol.*, 3 (1973) pp. 589–97.

13 T. L. W. Rothwell, B. M. Ogilvie and R. J. Love, 'Acetylcholinesterase Secretion by Parasitic Nematodes. II. *Trichostrongylus* spp.', *Int. J. Parasitol.*, 3 (1973) pp. 598–608.

14 D. McLaren, 'Nematode Sense Organs', *Adv. Parasitol.*, 14 (1976) pp. 195–265.

15 A. O. Anya, 'Physiological Aspects of Reproduction in Nematodes', *Adv. Parasitol.*, 14 (1976) pp. 267–351.

16 C. D. Green, 'Mating and Host Finding Behaviour of Plant Nematodes', in *Plant Parasitic Nematodes,* Vol. 2, eds B. M. Zuckerman, W. F. Mai and R. A. Rhode (Academic Press, New York, 1971) pp. 247–66.

17 C. D. Green and D. N. Greet, 'The Location of the Secretions that Attract Male *Heterodera schachtii* and *H. rostochiensis* to their Females', *Nematologica,* 18 (1972) pp. 347–52.

18 A. C. Triantophyllou, 'Sex Determination in *Meloidogyne incognita* (Chitwood, 1949) and Intersexuality in *M. javanica*', *Ann. Inst. Phytopath, Benaki NS.*, 3 (1960) pp. 12–31.

19 A. C. Triantophyllou, 'Genetics and Cytology', in *Plant Parasitic Nematodes,* Vol. 2, eds B. M. Zuckerman, W. F. Mai and R. A. Rhode (Academic Press, New York, 1971).

20 P. L. Lippens, 'Ultrastructure of a Marine Nematode, *Chromadorina germanica* (Buetschli, 1874). II. Cytology of Lateral Epidermal Glands and Associated Neurocytes', *Z. Morph. Tiere,* 79 (1974) pp. 283–94.
21 K. A. Wright, and J. Chan, 'Sense Receptors in the Bacillary Band of the Trichuroid Nematodes', *Tissue and Cell,* 5 (1973) pp. 373–80.
22 P. L. Lippens, 'Ultrastructure of a Marine Nematode, *Chromadorina germanica* (Buetschli, 1874). I. Anatomy and Cytology of the Caudal Gland Apparatus', *Z. Morph. Tiere,* 78 (1974) pp. 181–92.
23 K. A. Wright and W. D. Hope, 'Elaborations of the Cuticle of *Acanthonchus duplicatus* (Wieser, 1959) (Nematoda: Cyatholaimidae) as Revealed by Light and Electron Microscopy', *Canad. J. Zool.,* 46 (1968) pp. 1005–11.
24 D. L. Lee, '*Nippostrongylus brasiliensis*: Some Aspects of the Fine Structure and Biology of the Infective Larva and Adult', in *Nippostrongylus and Toxoplasma. Symposia of the British Society for Parasitology* Vol. 7, ed. A. E. R. Taylor (Blackwell, Oxford, 1969) pp. 3–16.
25 D. L. Lee, 'Moulting in Nematodes: The Formation of the Adult Cuticle During the Final Mould of *Nippostrongylus brasiliensis*', *Tissue and Cell,* 2 (1970) pp. 139–53.
26 E. Wisse and W. T. Daems, 'Electron Microscopic Observations on Second-stage Larvae of the Potato Root Eelworm *Heterodera rostochiensis*', *J. Ultrastr. Res.,* 24 (1968) pp. 210–31.
27 A. M. Shepherd, S. A. Clark and P. J. Dart, 'Cuticle Structure in the Genus *Heterodera*', *Nematologica,* 18 (1972) pp. 1–17.
28 I. Riding, 'Microvilli on the Outside of a Nematode', *Nature,* 226 (1970) pp. 179–80.
29 S. Fuchs and W. F. Harrington, 'Immunological Properties of *Ascaris* Cuticle Collagen', *Biochim. Biophys. Acta.,* 221 (1970) pp. 119–24.
30 D. Fujimoto and S. Kanaya, 'Cuticlin: A Noncollagen Structural Protein from *Ascaris* Cuticle', *Arch. Biochem. Biophys.,* 157 (1973) pp. 1–6.
31 C. Ellenby and L. Smith, 'Influence of Temperature on Polyphenol Oxidase Activity in Three Species of Heterodera', *Comp. Biochem. Physiol.,* 21 (1967) pp. 51–7.
32 T. P. Bonner, P. P. Weinstein and H. Saz, 'Synthesis of Cuticular Protein During the Third Moult in the Nematode *Nippostrongylus brasiliensis*', *Comp. Biochem. Physiol.,* 40B (1971) pp. 121–7.
33 D. Fujimoto and D. J. Prokop, 'Protocollagen Proline Hydroxylase from *Ascaris lumbricoides*', *J. Biol. Chem.,* 244 (1969) pp. 205–10.
34 W. P. Rogers and R. I. Somerville, 'The Infectious Process and its Relation to the Development of Early Parasitic Stages of Nematodes', *Adv. Parasitol.,* 6 (1968) pp. 327–48.
35 W. P. Rogers and R. I. Sommerville, 'Chemical Aspects of Growth and Development', in *Chemical Zoology* Vol. 3, eds M. Florkin and B. T. Scheer (Academic Press, New York, 1969) pp. 465–99.
36 P. Grootaert and P. L. Lippens, 'Some Ultrastructural Changes in Cuticle and Hypodermis of *Aporcellaimellus* During the First Mould (Nematoda: Dorylaimoidea)', *Z. Morphol. Tiere,* 79 (1974) pp. 269–82.
37 D. R. Roggen, D. J. Raski and N. O. Jones, 'Further Electron Microscopic Observations of *Xiphinema Index*', *Nematologica,* 13 (1967) pp. 1–16.
38. P. W. Johnson, S. D. van Gundy and W. W. Thomson, 'Cuticle Formation in *Hemicycliophora arenaria, Aphelenchus avenae* and *Hirschmaniella gracillis*', *J. Nematol.,* 2 (1970) pp. 59–79.
39 B. L. Gunter and L. Kampfe, 'Bau und Veronderung des Integumentes im

Entwicklungszyklus Cystenbildender Nematoden', *Verh. dt. Zool. Ges.,* Year 1966 (1967) pp. 152–66.

40 M. R. Samoiloff and J. Pasternak, 'Nematode Morphogenesis: Fine Structure of the Moulting Cycles in *Panagrellus silusiae* (de Man, 1913; Goodey, 1945), *Canad. J. Zool.,* 47 (1969) pp. 639–43.

41 T. P. Bonner, M. G. Menefie and F. J. Etges, 'Ultrastructure of Cuticle Formation in a Parasitic Nematode', *Z. Zellforsch.,* 104 (1970) pp. 193–203.

42 W. J. Kozek, 'The Moulting Pattern in *Trichinella spiralis.* II. An Electron Microscope Study', *J. Parasitol.,* 57 (1971) pp. 1029–38.

43 K. G. Davey and S. P. Kan, 'Moulting in a Parasitic Nematode, *Phocanema decipens*–IV. Ecdysis and its Control', *Canad. J. Zool.,* 46 (1968) pp. 893–8.

44 K. G. Davey and R. I. Sommerville, 'Moulting in a Parasitic Nematode, *Phocanema decipens*–VII. The Mode of Action of the Ecdysial Hormone', *Int. Il. Parasitol,* 4 (1974) pp. 241–59.

45 W. P. Rogers, 'Juvenile and Moulting Hormone from Nematodes', *Parasitology,* 67 (1973) pp. 105–13.

46 K. G. Davey, 'Moulting in a Parasitic Nematode, *Phocanema decipens*–VI. The Mode of Action of Insect Juvenile Hormone and Farnesyl Methyl Ether', *Int. J. Parasitol.,* 1 (1971) pp. 61–6.

47 C. J. Mapes, 'Bile and Bile Salts and Exsheathment of the Intestinal Nematodes *Trichostrongylus colubriformis* and *Nematodirus battus*', *Int. J. Parasitol.,* 2 (1972) pp. 433–8.

48 W. P. Rogers, '*The Nature of Parasitism. The Relationship of some Metazoan Parasites to their Hosts*', (Academic Press, New York, and London, 1962).

49 J. O. D. Slocombe, 'Some Analyses of Exsheathing Fluid from Infective *Haemonchus contortus* Larvae from Ontario', *Int. J. Parasitol.,* 4 (1974) pp. 397–402.

50 H. R. Wallace, *Nematode Ecology and Plant Disease* (Edward Arnold, London, 1973) p. 228.

51 V. H. Dropkin, 'Cellular Responses of Plants to Nematode Infections', *Ann. Rev. Phytopath.,* 7 (1969) pp. 101–22.

52 K. A. Wright, 'The Feeding Site and Probable Feeding Mechanism of the Parasitic Nematode *Capillaria hepatica* (Bancroft, 1893), *Canad. J. Zool.,* 52 (1974) pp. 1215–20.

53 J. B. Colam, 'Studies on Gut Ultrastructure and Digestive Physiology in *Cosmocerca ornata* (Nematoda: Ascaridida), *Parasitology,* 62 (1971) pp. 259–72.

54 C. J. Mapes, 'Structure and Function in the Nematode Pharynx. I. The Structure of the Pharynx of *Ascaris lumbricoides, Oxyuris equi, Aplectana brevicaudata* and *Panagrellus silusiae*', *Parasitology,* 55 (1965) pp. 269–84.

55 D. R. Roggen, 'Striated Muscles in the Pharynx of *Enoplus communis*', *Nematologica,* 16 (1970) pp. 603–5.

56 R. H. Waterstow, H. F. Epstein and S. Brenner, 'Paramyosin of *Caenorhabditis elegans*', *J. molec. Biol.,* 90 (1974) pp. 285–90.

57 R. G. Bruce, 'Structure of the Esophagus of the Infective Juvenile and Adult *Trichinella spiralis*', *J. Parasitol.,* 56 (1970) pp. 540–9.

58 K. A. Wright, 'The fine Structure of the Esophagus of some Trichuroid Nematodes. II. The Buccal Capsule and Anterior Esophagus of *Capillaria hepatica* (Bancroft, 1893)', *Canad. J. Zool.,* 52 (1974) pp. 1207–13.

59 S. Ward, N. Thomson, J. G. White and S. Brenner, 'Electron Microscopical Reconstruction of the Anterior Sensory Anatomy of the Nematode *Caenorhabditis elegans*', *J. Comp. Neurol.,* 160 (1975) pp. 313–38.

60 J. B. Jennings and J. B. Colam, 'Gut Structure, Digestive Physiology and Food Storage in *Pontonema vulgaris* (Nematoda: Enoplida)', *J. Zool.*, 161 (1970) pp. 211–21.

61 D. R. Roggen, 'Functional Morphology of the Nematode Pharynx. I. Theory of the Soft-walled Cylindrical Pharynx, *Nematologica*, 19 (1973) pp. 349–65.

62 C. C. Doncaster, 'Feeding in Plant Parasitic Nematodes: Mechanisms and Behaviour', in *Plant Parasitic Nematodes*, Vol. 2, eds B. M. Zuckerman, W. F. Mai and R. A. Rohde (Academic Press, New York, 1971) pp. 137–57.

63 M. K. Seymour and C. C. Doncaster, 'Defecation Behaviour of *Aphelenchoides blastophthorus*', *Nematologica*, 28 (1973) pp. 463–8.

64 M. K. Seymour and A. M. Shepherd, 'Cell Junctions Acting as Intestinal Valves in Nematodes', *J. Zool.*, 173 (1974) pp. 517–23.

65 N. A. Croll, 'Energy Utilisation of Infective *Ancylostoma tubaeforme* Larvae', *Parasitology*, 64 (1972) pp. 355–68.

66 C. C. Doncaster and M. K. Seymour, 'Exploration and Selection of Penetration Site by Tylenchida', *Nematologica*, 19 (1973) pp. 137–45.

67 M. K. Seymour, 'Defecation in a Passively Feeding Plant Nematode, *Hexotylus viviparus* (Neotylenchidae: Tylenchida)', *Nematologica*, 20 (1974) pp. 355–60.

68 C. C. Doncaster and M. K. Seymour, 'Passive Ingestion in a Plant Nematode *Hexatylus viviparus.* (Neotylenchidae: Tylenchida.)', *Nematologica*, 20 (1974) pp. 297–307.

69 F. D. McElroy and S. D. van Gundy, 'Observations on the Feeding Processes of *Hemicycliophora arenaria*', *Phytopathology*, 58 (1968) pp. 1558–65.

70 A. F. Bird, 'Symbiotic Relationships Between Nematodes and Plants', in *Symbiosis. Symposia of the Society for Experimental Biology*, Vol. 29, eds D. H. Jennings and D. L. Lee (Cambridge University Press, 1975) pp. 351–71.

71 H. C. Hechler, 'Description, Developmental Biology and Feeding Habits of *Seinura tenuicaudata* (De Man), J. B. Goodey, 1960 (Nematoda: Aphelenchoididae), a Nematode Predator', *Proc. Helminthol. Soc. Wash.*, 30 (1963) pp. 182–95.

72 V. H. Dropkin, 'Physiology of Nematodes of the Soil', *Ann. N.Y. Acad. Sci.*, 139 (1966) pp. 39–52.

73 R. Gordon and J. M. Webster, 'Biological Control of Insects by Nematodes', *Helminth. Abs.*, 43A (1974) pp. 327–49.

74 J. B. Colam, 'Studies on Gut Ultrastructure and Digestive Physiology in *Cyathostoma lari* (Nematoda: Strongylida)', *Parasitology*, 62 (1971) pp. 273–83.

75 U. P. Kolkofen, 'Attachment and Feeding Behaviour of *Ancylostoma caninum*', *Z. Parasit.*, 33 (1970) pp. 339–54.

76 D. Orion and R. Bronner, 'The Localization of Starch, Amylase and Invertase in *Meloidogyne javanica* Galls', *Nematologica*, 19 (1973) pp. 401–2.

77 D. L. Lee, 'Penetration of Mammalian Skin by the Infective Larva of *Nippostrongylus brasiliensis*', *Parasitology*, 65 (1972) pp. 499–505.

78 B. E. Matthews, 'Mechanism of Skin Penetration by *Ancylostoma tubaeforme* Larvae', *Parasitology*, 70 (1975) pp. 25–38.

79 J. B. Colam, 'Studies on Gut Ultrastructure and Digestive Physiology in *Rhabdias bufonis* and *R. sphaerocephala* (Nematoda: Rhabditida)', *Parasitology*, 62 (1971) pp. 247–58.

80 R. F. Myers, 'Amylase, Cellulase, Invertase and Pectinase in several Free-Living, Mycophagous and Plant-parasitic nematodes', *Nematologica*, 11 (1965) pp. 441–8.

81 P. Leštan, 'Chitin and Chitinolysis in Ascaridae', *Biologia, Bratisl.*, 24 (1969) pp. 601–6.
82 K. H. Ward and D. Fairbairn, 'Chitinase in Developing Eggs of *Ascaris suum* (Nematoda)', *J. Parasitol.*, 58 (1972) pp. 546–9.
83 D. L. Lee and M. H. Smith, 'Haemoglobins of Parasitic Animals', *Exp. Parasitol.*, 16 (1965) pp. 392–424.
84 J. Riley, 'Histochemical and Ultrastructural Observations on Digestion in *Tetrameres fissispina* (Diesing, 1861) (Nematoda: Spiruidea) with Special Reference to Intra-cellular Digestion', *Int. J. Parasitol.*, 3 (1973) pp. 157–64.
85 D. L. Lee, 'Structural and Biochemical Changes in *Nippostrongylus brasiliensis* During Development of Immunity to this Nematode in Rats', in *Comparative Biochemistry of Parasites*, ed. H. van den Bossche (Academic Press, New York, 1972) pp. 317–21.
86 G. K. Reznik, 'The Intestinal Secretion of the Parasitic Nematode of Domestic Fowl *Ascaridia galli*', *Doklady Akademii Nauk SSSR*, 182 (1968) pp. 735–6.
87 R. Gordon and J. M. Webster, 'Nutritional Requirements for Protein Synthesis During Parasitic Development of the Entomophilic Nematode *Mermis nigrescens*', *Parasitology*, 64 (1972) pp. 161–72.
88 T. G. Rutherford and J. M. Webster, 'Transcuticular Uptake of Glucose by the Entomophilic Nematode *Mermis nigrescens*', *J. Parasitol.*, 60 (1974) pp. 804–8.
89 R. P. Harpur and J. S. Popkin, 'Intestinal Fluid Transport: Studies with the Gut of *Ascaris lumbricoides*', *Can. J. Physiol. Pharmacol.*, 51 (1973) pp. 79–90.
90 L. M. Schanbacher and C. G. Beames, 'Transport of 3-O-Methylglucose by Various Regions of the Intestine of *Ascaris suum*', *J. Parasitol.*, 59 (1973)
91 C. G. Beames, H. H. Bailey, C. O. Rock and L. M. Schanbacher, 'Movement of Cholesterol and β-sitosterol across the Intestine of *Ascaris suum*', *Comp. Biochem. Physiol.*, 47A (1974) pp. 881–8.
92 W. L. Nicholas, *The Biology of Free-living Nematodes* (Clarendon Press, Oxford, 1975).
93 P. P. Weinstein, 'The *In Vitro* Cultivation of Helminths with Reference to Morphogenesis', in *Biology of Parasites*, ed. E. J. L. Soulsby (Academic Press, New York, 1966) pp. 143–54.
94 H. Keilová and V. Tomášek, 'Effect of Pepsin Inhibit from *Ascaris lumbricoides* on Cathepsin D and E,' *Biochim, Biophys. Acta*, 284 (1972) pp. 461–4.
95 A. F. Cooper and S. D. van Gundy, 'Ethanol Production and Utilization by *Aphelenchus avenae* and *Caenorhabditis* sp.', *J. Nematol.*, 3 (1971) pp. 205–14.
96 L. R. Krusberg, 'Chemical Composition of Nematodes', in *Plant Parasitic Nematodes*, Vol. 2, eds B. M. Zuckerman, W. F. Mai and R. A. Rohde (Academic Press, New York, 1971) pp. 213–34.
97 J. Barrett, C. W. Ward and D. Fairbairn, 'The Glyoxylate Cycle and the Conversion of Triglycerides to Carbohydrates in Developing Eggs of *Ascaris lumbricoides*', *Comp. Biochem. Physiol.*, 35 (1970) pp. 577–86.
98 J. D. Ferguson and G. A. Castro, 'Metabolism of Intestinal Stages of *Trichinella spiralis*', *Amer. J. Physiol.*, 225 (1973) pp. 85–9.
99 A. F. Cooper and S. D. van Gundy, 'Metabolism of Glycogen and Neutral Lipids by *Aphelenchus avenae* and *Caenorhabditis* sp. in Aerobic, Microaerobic and Anaerobic Environments', *J. Nematol.*, 2 (1970) pp. 305–15.

100 J. Barrett, C. W. Ward and D. Fairbairn, 'Lipid Metabolism in the Free-living Nematodes *Panagrellus redivivus* and *Turbatrix aceti*', *Comp. Biochem. Physiol.*, 38B (1971) pp. 279–84.
101 G. E. Tarr, 'Comparison of Ascarosides from Four Species of Ascarid and One of Oxyurid (Nematoda)', *Comp. Biochem. Physiol.*, 46B (1973) pp. 167–76.
102 G. E. Tarr and D. Fairbairn, 'Conversion of Ascaroside Esters to Free Ascarosides in Fertilised Eggs of *Ascaris suum* (Nematoda)', *J. Parasitol.*, 59 (1973) pp. 428–33.
103 D. Fairbairn, 'Biochemical Adaptation and Loss of Genetic Capacity in Helminth Parasites', *Biol. Rev.*, 45 (1970) pp. 29–72.
104 J. Barrett and I. Beis, 'Studies on Glycolysis in the Muscle Tissue of *Ascaris lumbricoides* (Nematoda)', *Comp. Biochem. Physiol.*, 44B (1973) pp. 751–61.
105 C. Bryant, 'Carbon Dioxide Utilisation and the Regulation of Respiratory Metabolic Pathways in Parasitic Helminths', *Adv. Parasitol.*, 13 (1975) pp. 35–69.
106 G. W. Hutchinson and M. A. Fernando, 'Enzymes of Glycolysis and the Pentose Phosphate Pathway During Development of the Rabbit Stomach Worm *Obeliscoides cuniculi*', *Int. J. Parasitol.*, 4 (1974) pp. 389–95.
107 H. van den Bossche, O. F. J. van Parijs and D. Thienpont, 'Studies on the Carbohydrate Metabolism of Third-stage *Haemonchus contortus* Larvae', *Life Sci.*, 8 (1969) pp. 1047–54.
108 H. J. Saz, 'Comparative Biochemistry of Carbohydrates in Nematodes and Cestodes', in *Comparative Biochemistry of Parasites*, ed. H. van den Bossche (Academic Press, New York, 1972) pp. 33–47.
109 D. Fairbairn, 'The Physiology and Biochemistry of Nematodes', in *Nematology. Fundamentals and Recent Advances with Emphasis on Plant Parasitic and Soil Forms*, eds J. N. Sasser and W. R. Jenkins (University of North Carolina Press, Chapel Hill, 1960).
110 W. Körting and D. Fairbairn, 'Changes in Beta-oxidation and Related Enzymes During the Life Cycle of *Strongyloides ratti* (Nematoda)', *J. Parasitol.*, 57 (1971) pp. 1029–38.
111 H. J. Atkinson, 'The Respiratory Physiology of the Marine Nematodes *Enoplus brevis* (Bastian) and *E. communis*. I. The Influence of Oxygen Tension and Body Size', *J. exp. Biol.*, 59 (1973) pp. 255–66.
112 H. J. Atkinson, 'The Respiratory Physiology of the Marine Nematodes *Enoplus brevis* (Bastian) and *E. communis* (Bastian). II. The Effects of Changes in the Imposed Oxygen Regime', *J. exp. Biol.*, 59 (1973) pp. 267–74.
113 T. J. W. Alphey, 'An *In Vitro* Study of the Effect of Oxygen Tension upon the Motility of *Nippostrongylus brasiliensis*', *Parasitology*, 64 (1972) pp. 181–6.
114 J. M. Castillo and L. R. Krusberg, 'Organic Acids of *Ditylenchus triformis* and *Turbatrix aceti*', *J. Nematol.*, 3 (1971) pp. 284–8.
115 K. M. McNeill and W. F. Hutchinson, 'The Tricarboxylic Acid Cycle Enzymes in the Adult Dog Heartworm *Dirofilaria immitis*', *Comp. Biochem. Physiol.*, 38B (1971) pp. 493–500.
116 M. Rothstein and H. Mayoh, 'Nematode Biochemistry – VIII. Malate Synthetase', *Comp. Biochem. Physiol.*, 17 (1966) pp. 1181–8.
117 M. Rothstein, F. Nicholls and P. Nicholls, 'Nematode Biochemistry. XII. Mitochondria from the Free-living Nematode *Turbatrix aceti*', *Int. J.*

Biochem., 1 (1970) pp. 695–705.
118 W. J. Vaastra, 'Intermediary Metabolism of the Cattle Lungworm *Dictyocaulus viviparus*', *Hoppe-Seyler's Z. physiol. Chem.*, 350 (1969) pp. 701–9.
119 B. Chance, B. Schoener and F. Schindler, 'The Intracellular Oxidation–Reduction State', in *Oxygen in the Animal Organism. I.U.B. Symposium 31*, eds F. Dickens and E. Neil (Pergamon Press, Oxford, 1964).
120 D. W. T. Crompton, D. H. Shrimpton and I. A. Silver, 'Measurements of the Oxygen Tension in the Lumen of the Small Intestine of the Domestic Duck', *J. exp. Biol.*, 43 (1965) pp. 473–8.
121 E. J. Wang and H. J. Saz, 'Comparative Biochemical Studies of *Litomosoides carinii, Dipetalonema viteae* and *Brugia pahangi* Adults', *J. Parasitol.*, 60 (1974) pp. 316–21.
122 K. S. Cheah, 'Properties of *Ascaris* Muscle Mitochondria. I. Cytochromes', *Biochim. Biophys. Acta,* 387 (1975) pp. 107–14.
123 L. C. Costello, W. N. Smith and W. Fredericks, 'The Comparative Biochemistry of Developing *Ascaris* Eggs. VI. Respiration and Terminal Oxidation During Embryonation', *Comp. Biochem. Physiol.*, 18 (1966) pp. 217–24.
124 H. J. Saz, 'Effects of Anthelmintics on ^{32}P Esterification in Helminth Metabolism', in *Comparative Biochemistry of Parasites,* ed. H. van den Bossche (Academic Press, New York, 1972) pp. 445–54.
125 L. G. Warren, 'Biochemistry of the Dog Hookworm. III. Oxidative Phosphorylation', *Exp. Parasitol.*, 27 (1970) pp. 417–23.
126 L. R. Krusberg, 'Hydrolytic and Respiratory Enzymes of Species of *Ditylenchus* and *Pratylenchus*', *Phytopathology,* 50 (1960) pp. 9–22.
127 B. W. Langer, W. J. Smith and V. J. Theodorides, 'The Pentose Cycle in Adult *Ascaris suum*', *J. Parasitol.*, 57 (1971) pp. 485–6.
128 J. Barrett, 'Nucleoside Triphosphate Metabolism in the Muscle Tissue of *Ascaris lumbricoides* (Nematoda)', *Int. J. Parasitol.*, 3 (1973) pp. 393–400.
129 J. Barrett and I. Beis, 'Nicotinamide and Adenosine Nucleotide Levels in *Ascaris lumbricoides, Hymenolepis diminuta* and *Fasciola hepatica*', *Int. J. Parasitol.*, 3 (1973) pp. 271–3.
130 L. G. Warren and W. J. Poole, 'Biochemistry of the Dog Hookworm. II. Nature and Origin of the Excretory Fatty Acids', *Exp. Parasitol.*, 27 (1970) pp. 408–16.
131 J. Barrett, G. D. Cain and D. Fairbairn, 'Sterols in *Ascaris lumbricoides* (Nematoda), *Macrocanthorhynchus hirudinaceus* and *Moniliformis dubius* (Acanthocephala) and *Echinostoma revolutum* (Trematoda)', *J. Parasitol.*, 56 (1970) pp. 1004–8.
132 R. J. Cole and L. R. Krusberg, 'Sterol Metabolism in *Turbatrix aceti*', *Life Sci.*, 7 (1968) pp. 713–24.
133 M. Rothstein, 'Nematode Biochemistry. IX. Lack of Sterol Biosynthesis in Free-living Nematodes', *Comp. Biochem. Physiol.*, 27 (1968) pp. 308–17.
134 D. L. Lee, 'Changes in Adult *Nippostrongylus brasiliensis* During the Development of Immunity to this Nematode in Rats. 2. Total Lipids and Neutral Lipids', *Parasitology,* 63 (1971) pp. 271–4.
135 A. J. Clarke, P. M. Cox and A. M. Shepherd, 'The Chemical Composition of the Egg Shells of the Potato Cyst Nematode *Heterodera rostochiensis.* Woll.', *Biochem. J.*, 104 (1967) pp. 1056–60.
136 M. A. Fernando, 'Hemoglobins of Parasitic Nematodes. II. Electrophoretic Analysis of the Multiple Hemoglobins of Adults and Developmental Stages of the Rabbit Worm, *Obeliscoides cuniculi*', *J. Parasitol.*, 55 (1969) pp. 493–7.

137 T. Okazaki, B. A. Wittenberg, R. W. Briehl and J. B. Wittenberg, 'The Hemoglobin of *Ascaris* Body Wall', *Biochim. Biophys. Acta.*, 140 (1967) pp. 258–65.
138 M. E. Perez-Gimenez, A. Gimenez and K. Gaede, 'Metabolic Transformation of ^{14}C-glucose into Tissue Proteins of *Ancylostomac aninum', Exp. Parasitol.*, 21 (1967) pp. 215–23.
139 D. J. Wright and D. R. Newall, 'Nitrogen Excretion, Osmotic and Ionic Regulation in Nematodes', in *The Organization of Nematodes,* ed. N. A. Croll (Academic Press, New York and London, 1976).
140 B. W. Langer, 'The Glutamic Acid Dehydrogenase of Adult Female *Ascaris suum*', *J. Parasitol.,* 58 (1972) pp. 539–40.
141 G. O. Poinar and R. J. Hess, 'Food Uptake by the Insect-parasitic Nematode *Sphaerularia bombi* (Tylenchida)', *J. Nematol.,* 4 (1972) pp. 270–7.
142 C. W. Miller and R. N. Roberts, 'Alanine Synthesis in the Nematode *Aphelenchoides ritzemabosi', Phytopathology,* 54 (1964) p. 1177.
143 M. Chvapil, M. Boucek and E. Ehrlich, 'Differences in the Protocollagen Hydroxylase Activities from *Ascaris* Muscle and Hypodermis', *Arch. Biochem. Biophys.,* 140 (1970) pp. 11–18.
144 H. J. Atkinson, 'The Respiratory Physiology of Nematodes', in *The Organization of Nematodes,* ed. N. A. Croll (Academic Press, New York and London, 1976).
145 R. Z. Klewkowski, L. Wasilewska and E. Paplinska, 'Oxygen Consumption in the Developmental Stages of *Panagrolaimus rigidus', Nematologica,* 20 (1974) pp. 61–8.
146 V. Bryant, 'Growth and Respiration Throughout the Life Cycle of *Nematospiroides dubius* (Baylis, 1926) (Nematoda : Heligmosomidae). The Free Living Stages', *Parasitology,* 67 (1973) pp. 245–51.
147 B. D. Bhatt and R. A. Rhode, 'The Influence of Environmental Factors on the Respiration of Plant-parasitic Nematodes', *J. Nematol.,* 2 (1970) pp. 277–85.
148 E. K. Mercer and E. J. Cairns, 'Food Consumption of the Free-living Aquatic Nematode *Pelodera chitwoodi', J. Nematol.,* 5 (1973) pp. 201–8.
149 I. K. Barker, 'Scanning Electron Microscopy of the Duodenal Mucosa of Lambs Infected with *Trichostrongylus colubriformis*', *Parasitology,* 67 (1973) pp. 307–14.
150 M. H. Smith, 'The Pigments of Nematodes and Acanthocephala', in *Chemical Zoology* Vol. 3, eds M. Florkin and B. T. Scheer (Academic Press, New York, 1969) pp. 501–20.
151 R. I. Baxter and C. D. Blake, 'Oxygen and Hatch of Eggs and Migration of Larvae in *Meloidogyne javanica', Ann. appl. Biol.,* 63 (1969) pp. 191–203.
152 A. F. Cooper, S. D. van Gundy and L. H. Stolzy, 'Nematode Reproduction in Environments of Fluctuating Aeration', *J. Nematol.,* 2 (1970) pp. 182–8.
153 J. Klinger and P. Kunz, 'Investigations with a Saprozoic Nematode, *Mesodiplogaster lheritieri* on a Possible Respiratory Function of Air Swallowing', *Nematologica,* 20 (1974) pp. 52–60.
154 M. H. Smith and D. L. Lee, 'Metabolism of Haemoglobin and Haematin Compounds in *Ascaris lumbricoides*', *Proc roy. Soc.,* B. 157 (1963) pp. 234–57.
155 J. E. Rose and K. L. Kaplan, 'Purification, Molecular Weight, and Oxygen Equilibrium of Hemoglobin from *Syngamus trachea*, the Poultry Gapeworm', *J. Parasitol.,* 58 (1972) pp. 903–6.
156 J. B. Wittenberg, 'Myoglobin-facilitated Oxygen Diffusion: Role of Myo-

globin in Oxygen Entry into Muscle', *Physiol. Rev.,* 50 (1970) pp. 559–636.
157 H. J. Atkinson, 'The Functional Significance of the Haemoglobin in a Marine Nematode *Enoplus brevis* (Bastian)', *J. exp. Biol.,* 62 (1975) pp. 1–9.
158 J. del Castillo, 'Pharmacology of Nematoda', in *Chemical Zoology,* Vol. 3, ed. M. Florkin and B. T. Scheer (Academic Press, New York, 1969) pp. 521–4.
159 B. C. Sanderson, 'Anthelmintics in the Study of Helminth Metabolism', in *Chemotherapeutic Agents in the Study of Parasites. Symposia of the British Society for Parasitology,* Vol. 11, eds A. E. R. Taylor and R. Muller (Blackwell Scientific Publications, Oxford, 1973) pp. 53–82.
160 H. van den Bossche, *Comparative Biochemistry of Parasites* (Academic Press, New York, 1972).
161 S. V. Govorova and O. I. Polyakova, 'Effect of Benzimidazole Derivatives on the Enzymes of *Ascaridia galli*', *Trudy Vsesoyuznogo Instituta Gel'mintologu im K. I. Skryabina,* 18 (1971) pp. 75–7.
162 R. K. Pritchard, 'The Fumarate Reductase Reaction of *Haemonchus contortus* and the Mode of Action of Some Anthelmintics', *Int. J. Parasitol.,* 3 (1973) pp. 409–17.
163 C. E. Castro and I. J. Thomason, 'Mode of Action of Nematicides', in *Plant Parasitic Nematodes,* Vol. 2, eds B. M. Zuckerman, W. F. Mai and R. A. Rohde (Academic Press, New York, 1971) pp. 289–96.
164 C. L. Prosser, *Comparative Animal Physiology,* 3rd ed. (W. B. Saunders Co., Philadelphia, 1973).
165 D. R. Viglierchio, 'Osmoregulation and Electrolyte Uptake in Antartic Nematodes', *Trans. am. Micros. Soc.,* 93 (1974) pp. 325–38.
166 R. D. Winslow, 'Some Aspects of the Ecology of the Free-living and Plant-parasitic Nematodes', in *Nematology. Fundamentals and Recent Advances with Emphasis on Plant Parasitic and Soil Forms,* eds J. N. Sasser and W. R. Jenkins (University of North Carolina Press, Chapel Hill, 1960).
167 H. R. Wallace 'Abiotic Influences in the Soil Environment', in *Plant Parasitic Nematodes,* Vol. 1, eds B. M. Zuckerman, W. F. Mai and R. A. Rohde (Academic Press, New York, 1971) pp. 257–80.
168 W. Stephenson, 'The Effect of Variations in Osmotic Pressure upon a Free-living Soil Nematode', *Parasitology,* 34 (1942) pp. 253–65.
169 D. L. Lee, 'The Effect of Changes in the Osmotic Pressure upon *Hammerschmidtiella diesingi* (Hammerschmidt, 1838) with Reference to the Survival of the Nematode During Moulting of the Cockroach', *Parasitology,* 50 (1960) pp. 241–6.
170 A. O. Anya, 'Investigations on Osmotic Regulation in the Parasitic Nematode *Aspiculuris tetraptera* Shulz', *Parasitology,* 56 (1966) pp. 583–8.
171 A. D. Hobson, W. Stephenson and L. C. Beadle, 'Studies on the Physiology of *Ascaris lumbricoides.* 1. The Relation of the Total Osmotic Pressure, Conductivity and Chloride Content of the Body Fluid to that of the External Environment', *J. exp. Biol.,* 29 (1952) pp. 1–21.
172 C. F. Marks, I. J. Thomason and C. E. Castro, 'Dynamics of the Permeation of Nematodes by Water, Nematocides and other Substances', *Exp. Parasitol.,* 22 (1968) pp. 321–37.
173 R. P. Harpur and J. S. Popkin, 'Osmolarity of Blood and Intestinal Contents in the Pig, Guinea Pig and *Ascaris lumbricoides*', *Can. J. Biochem.,* 43 (1965) pp. 1157–69.
174 C. Ellenby, 'Dormancy and Survival in Nematodes', in *Symposia of the Society for Experimental Biology,* Vol. 23 (Cambridge University Press,

1969) pp. 83–97.
175 R. F. Myers, 'Osmoregulation in *Panagrellus redivivus* and *Aphelenchus avenae*', *Nematologica*, 12 (1966) pp. 579–86.
176 G. Kummel, L. Dankwarth, G. Braun-Schubert and K. H. Gertz, 'Zur Struktur und Funktion des Exkretionssystems von *Ascaris lumbricoides* L.', *Z. vergl. Physiol.*, 84 (1969) pp. 118–34.
177 G. N. J. le Patourel and D. J. Wright, 'Studies on the Mechanisms of Accumulation of Phorate by the Free-living Nematode *Panagrellus redivivus*', *Pestic. Biochem. Physiol.*, 4 (1974) pp. 144–52.
178 E. J. Arthur and R. C. Sanborn, 'Osmotic and Ionic Regulation in Nematodes', in *Chemical Zoology*, vol. 3, ed. M. Florkin and B. T. Scheer (Academic Press, New York and London, 1969) pp. 429–64.
179 A. A. Castro and D. Fairbairn, 'Comparison of Cuticular and Intestinal Absorption of Glucose by Adult *Ascaris lumbricoides*', *J. Parasitol.*, 55 (1969) pp. 13–16.
180 B. G. Harris, J. M. Talent and D. Fodge, 'Amino Acid Transport in Reproductive and Muscle Tissues of *Ascaris suum*', *J. Parasitol.*, 58 (1972) pp. 541–5.
181 M. Rothstein, 'Nitrogen Metabolism in Aschelminthes', in *Comparative Biochemistry of Nitrogen Metabolism*, Vol. 1, ed. J. W. Campbell (Academic Press, New York, 1970) pp. 91–102.
182 J. K. Pollack and D. Fairbairn, 'The Metabolism of *Ascaris lumbricoides* Ovaries. II. Amino Acid Transport', *Can. J. Biochem. Physiol.*, 33 (1955) pp. 307–16.
183 J. Barrett and I. Beis, 'The Redox State of the Free Nicotinamide–Adenine Dinucleotide Couple in the Cytoplasm and Mitochondria of Muscle Tissue from *Ascaris lumbricoides* (Nematoda)', *Comp. Biochem. Physiol.*, 44A (1973) pp. 331–40.
184 R. W. Paltridge and P. A. Janssens, 'A Reinvestigation of the Status of the Ornithine–Urea Cycle in the Adult *Ascaris lumbricoides*', *Comp. Biochem. Physiol.*, 40B (1971) pp. 503–13.
185 D. J. Wright, 'Studies on Nitrogen Catabolism in *Panagrellus redivivus* (Goodey, 1945) (Nematoda; Cephalobidae)', *Comp. Biochem. Physiol.*, 52B (1975) pp. 255–60.
186 W. P. Rogers, 'Nitrogenous components and their metabolism; Acanihocephula and Nematoda', in *Chemical Zoology*, vol. 3, ed. M. Florkin and B. T. Scheer (Academic Press, New York and London, 1969) pp. 379–428.
187 D. J. Wright, 'Elimination of Nitrogenous Compounds by *Panagrellus redivivus* (Goodey, 1945) (Nematoda: Cephalobidae)', *Comp. Biochem. Physiol.*, 52B (1975) pp. 247–53.
188 P. P. Weinstein, 'Excretory Mechanisms and Excretory Products of Nematodes: An Appraisal', in *Host Influence on Parasite Physiology*, ed. L. A. Stauber (Rutgers University Press, New Brunswick, 1960).
189 H. D. Crofton, *Nematodes* (Hutchinson University Library, 1966).
190 K. W. Behernz, 'Vergleichende Physiologische Untersuchungen über die Exkretion Parasitischer Nematoden mit Hilfe der Flupreszenmikroskopie', *Z. wiss. Zool.*, 159 (1956) pp. 129–64.
191 S. A. Clark, A. M. Shepherd and A. Kempton, 'Spicule Structure in some *Heterodera* spp.', *Nematologica*, 18 (1973) pp. 242–7.
192 W. E. Foor, 'Spermatozoon Morphology and Zygote Formation in Nematodes', *Biol. Reprod.*, Suppl. 2 (1970) pp. 177–202.
193 C. D. Green, Orientation of Male *Heterodera rostochiensis* Woll. and *H.*

schachtii Schm to their Females', *Ann. App. Biol.*, 58 (1966) pp. 327–39.

194 C. C. Doncaster and M. K. Seymour, 'Egg-transport in the Genital Tract and Egg-laying by *Aphelenchoides blastophthorus*', *Nematologica*, 19 (1973) pp. 379–88.

195 P. A. G. Wilson, 'The Effect of Weak Electrolyte Solutions on the Hatching Rate of the Eggs of *Trichostrongylus retortaeformis* (Zeder) and its Interpretation in Terms of a Proposed Hatching Mechanism of Strongyloid Eggs', *J. exp. Biol.*, 35 (1958) pp. 584–601.

196 C. C. Doncaster and A. M. Shepherd, 'The Behaviour of Second-stage *Heterodera rostochiensis* Larvae Leading to their Emergence from the Egg', *Nematologica*, 13 (1967) pp. 476–8.

197 C. Ellenby, 'Water Uptake and Hatching in the Potato Cyst Nematode *Heterodera rostochiensis* and the Root Knot Nematode, *Meloidogyne incognita*', *J. exp. Biol.*, 61 (1974) pp. 773–9.

198 H. R. Wallace, *The Biology of Plant Parasitic Nematodes* (Edward Arnold, London, 1963).

199 R. J. Thomas, 'The Role of Climate in the Epidemiology of Nematode Parasitism in Ruminants', in *The Effects of Meteorological Factors upon Parasites. Symposia of the British Society for Parasitology,* Vol. 12, ed. A. E. R. Taylor and R. Muller (Blackwell Scientific Publications, Oxford, 1974).

200 A. A. F. Evans, 'The Effect of Climate on Plant and Soil Nematodes', in *The Effects of Meteorological Factors upon Parasites. Symposia of the British Society for Parasitology,* Vol. 12, eds A. E. R. Taylor and R. Muller (Blackwell Scientific Publications, Oxford, 1974).

201 A. M. Shepherd and A. J. Clarke, 'Moulting and Hatching Stimuli', in *Plant Parasitic Nematodes,* Vol. 2, eds B. M. Zuckerman, W. F. Mai and R. A. Rhode (Academic Press, New York and London, 1971) pp. 267–87.

202 F. G. W. Jones and M. G. Jones, *Pests of Field Crops,* 2nd ed. (Arnold, London, 1974).

203 H. J. Atkinson and A. J. Ballantyne, 'Changes in Oxygen Consumption of *Heterodera rostochiensis* Cysts During Emergence of Juveniles After Stimulation by Potato Root Diffusate', (submitted for publication).

204 H. J. Atkinson and A. J. Ballantyne, 'Changes in the Adenylate Energy Charge of *Heterodera rostochiensis* Juveniles which have been Stimulated to Hatch by Potato Root Diffusate', (submitted for publication).

205 I. Beis and J. Barrett, 'Energy Metabolism in Developing *Ascaris lumbricoides* Eggs. II. The Steady State Content of Intermediary Metabolites', *Devl Biol.,* 42 (1975) pp. 188–95.

206 D. Fairbairn, 'Physiological Aspects of Egg Hatching and Larval Exsheathment in Nematodes', in *Host Influence on Parasite Physiology,* ed. L. A. Stauber (Rutgers University Press, New Brunswick, 1960).

207 W. D. Hope, 'Fine Structure of the Somatic Muscles of the Free-living Marine Nematode *Deontostoma californicum* (Steiner and Albin, 1933) (Leptosomotidae)', *Proc. Helminthol. Soc. Wash.*, 36 (1969) pp. 10–29.

208 J. Rosenbluth, 'Ultrastructure of Somatic Muscle Cells in *Ascaris lumbricoides.* II. Intermuscular Junctions, Neuromuscular Junction; and Glycogen Stores', *J. Cell Biol.,* 26 (1965) pp. 579–91.

209 H. Hirumi, D. J. Raski and N. O. Jones, 'Primitive Muscle Cells of Nematodes: Morphological Aspects of Platymyarian and Shallow Coelomyarian Muscles in Two Plant Parasitic Nematodes, *Trichodorus christiei* and *Longidorus elongatus*', *J. Ultrastr. Res.,* 34 (1971) pp. 517–43.

210 J. Rosenbluth, 'Obliquely Striated Muscle. III. Contraction Mechanism of

Ascaris Body Muscle', *J. Cell Biol.,* 34 (1967) pp. 15–33.
211 K. A. Wright, 'Cephalic Sense Organs of the Rat Hookworm, *Nippostrongylus brasiliensis*– Form and Function', *Canad. J. Zool.,* 53 (1975) pp. 1131–46.
212 J. Sulston, M. Dew and S. Brenner, 'Dopaminergic Neurons in the Nematode *Caenorhabditis elegans*', *J. comp. Neurol.,* 163 (1975) pp. 215–26.
213 A. J. Nelmes, 'Behavioural Responses of Larvae of *Heterodera rostochiensis* to Aldicarb and its Sulfoxide and Sulfone', *J. Nematol.,* 2 (1970) pp. 223–7.
214 G. N. le Patourel and D. J. Wright, 'Uptake and Metabolism of Phorate by the Free-living Nematode *Panagrellus redivivus*', *Pestic. Biochem. Physiol.,* 4 (1974) pp. 135–43.
215 H. W. Spurr, 'Nematode Cholinesterases and their Inhibition by Nematicidal Carbamates', *Phytopathology,* 56 (1966) pp. 902–3.
216 N. K. Pannikar and N. G. Sproston, 'Osmotic Relations of some Metazoan Parasites', *Parasitology,* 33 (1941) pp. 214–23.
217 M. K. Seymour, 'Motion and the Skeleton in Small Nematodes', *Nematologica,* 19 (1973) pp. 43–8.
218 H. R. Wallace, 'Dynamics of Nematode Movement', *Ann. Rev. Phytopath.,* 6 (1968) pp. 91–114.
219 J. Gray and H. W. Lissman, 'The Locomotion of Nematodes', *J. exp. Biol.,* 41 (1964) pp. 135–54.
220 J. Gray, 'Undulatory Propulsion', *Q. Jl microscoc. Sci.,* 94 (1953) pp. 857–71.
221 H. R. Wallace, 'Undulatory Locomotion of the Plant Parasitic Nematode *Meloidogyne javanica*', *Parasitology,* 58 (1968) pp. 377–91.
222 H. D. Crofton, 'The Vertical Migration of Infective Larvae of Strongyloid Nematodes', *J. Helminth.,* 28 (1954) pp. 35–52.
223 F. G. W. Jones, 'The Soil as an Environment for Plant Parasitic Nematodes', *Ann. app. Biol.,* 79 (1975) pp. 113–39.
224 F. G. W. Jones and D. M. Parrott, 'The Influence of Soil Structure and Moisture on Nematodes Especially *Xiphinema, Longidorus, Trichodorus* and *Heterodera* spp.', *Soil Biochem. Soil Biol.,* 1 (1969) pp. 153–65.
225 H. R. Wallace, 'The Movement of Nematodes in the External Environment', in *Ecology and Physiology of Parasites*, ed. A. H. Fallis (University of Toronto Press, 1971).
226 D. L. Lee, 'Two- and Three- dimensional Locomotion in the Nematode *Nippostrongylus brasiliensis*', in preparation.
227 E. M. Reed and H. R. Wallace, 'Leaping Locomotion by an Insect-parasitic Nematode', *Nature,* 206 (1965) pp. 210–11.
228 H. T. Streu, W. R. Jenkins and M. T. Hutchinson, 'Nematodes Associated with Carnations', *New Jersey Agric. Exp. Sta. Rugers Bull.,* 800 (1961).
229 H. Stauffer, 'Die Lokomotion der Nematoden', *Zool. Jb.,* 49 (1924) pp. 1–118.
230 J. Robinson, '*Pilobolus* spp. and the Translation of Infective Larvae of *Dictyocaulus viviparus* from Faeces to Pasture', *Nature,* 193 (1962) pp. 353–4.
231 P. Bovien, 'Some Types of Association Between Nematodes and Insects', *Vidensk. Medd. dansk. naturh. Foren. Kbh.,* 101 (1937).
232 E. R. Noble and G. A. Noble, *Parasitology. The Biology of Animal Parasites.* 3rd ed. (Lea & Febiger, Philadelphia, 1971).
233 N. A. Croll and J. M. Smith, 'Nematode Setae as Mechanoreceptors', *Nematologica,* 20 (1974) pp. 291–6.
234 S. Ward, 'Chemotaxis by the Nematode *Caenorhabditis elegans:* Identifica-

tion of Attractants and Analysis of the Response by Use of Mutants', *Proc. Nat. Acad. Sci. U.S.A.*, 70 (1973) pp. 817–21.

235 K. A. Wright, 'Cephalic Sense Organs of the Parasitic Nematode *Capillaria hepatica* (Bancroft, 1893)', *Canad. J. Zool.*, 52 (1974) pp. 1207–13.

236 N. A. Croll, A. A. F. Evans and J. M. Smith, 'Comparative Nematode Photoreceptors', *Comp. Biochem. Physiol.*, 51A (1975) pp. 139–43.

237 N. A. Croll, 'Behavioural Analysis of Nematode Movement', *Adv. Parasitol.*, 13 (1975) pp. 71–122.

238 N. A. Croll, 'Behaviour of Larval Nematodes', in *Behavioural Aspects of Parasite Transmission,* ed. E. U. Canning and C. A. Wright (Academic Press, London, 1972) pp. 30–52.

239 D. B. Dusembery, 'Analysis of Chemotaxis in the Nematode *Caenorhabditis elegans* by Countercurrent Separation', *J. exp. Zool.*, 188 (1974) pp. 41–7.

240 N. A. Croll, *The Behaviour of Nematodes* (Edward Arnold, London, 1970).

241 T. J. W. Alphey, 'Studies on the Aggregation Behaviour of *Nippostrongylus brasiliensis*', *Parasitology,* 63 (1971) pp. 109–17.

242 F. D. Griffin, 'Effect of Acclimation Temperature on Infection of Alfalfa by *Ditylenchus dipsaci*', *J. Nematol.*, 6 (1974) pp. 57–9.

243 J. Klinger, 'The Reaction of *Aphelenchoides fragariae* to Slit-like Micro-openings and to Stomatal Diffusion Gases', *Nematologica,* 16 (1970) pp. 417–22.

244 R. S. Pitcher, 'The Host–Parasite Relations and Ecology of *Trichodorus viruliferus* on Apple Roots, as Observed from an Underground Laboratory', *Nematologica,* 13 (1967) pp. 547–57.

245 T. J. W. Alphey, 'Studies on the Distribution and Site Location of *Nippostrongylus brasiliensis* within the Small Intestine of Laboratory Rats', *Parasitology,* 61 (1970) pp. 449–60.

246 D. L. Lee, 'Changes in adult *Nippostrongylus brasiliensis* during the development of immunity to this nematode in rats. I. Changes in ultrastructure', *Parasitology,* 59 (1969) pp. 29–39.

Author Index

The page number for each entry is followed by the reference number in italics.

Alphey, T. J. W. 64, 65, *113*; 172, *241*; 172, 178, *245*
Anya, A. O. 9, 10, 116, 118, 119, 120, 121, 122, 123, *15*; 102, *170*
Arthur, E. J. 107, *178*
Atkinson, H. J. 64, 88, *111*; 64, *112*; 86, 87, 89, 90, 91, 93, *144*; 91, 92, 93, 94, *157*; 130, 131, *203*; 130, 131, *204*

Bailey, H. H. 55, *91*
Ballantyne, A. J. 130, 131, *203*; 130, 131, *204*
Barker, I. K. 89, 155, *149*
Barrett, J. 58, 75, 78, 80, *97*; 59, 65, 75, *100*; 62, 63, 67, *104*; 76, *128*; 76, *129*; 79, *131*; 109, *183*; 78, 132, *205*
Baxter, R. I. 89, 127, *151*
Beadle, L. C. 102, 105, 106, 108, 113, *171*
Beames, C. G. 54, *90*; 55, 84, *91*
Behernz, K. W. 114, *190*
Beis, I. 62, 63, 67, *104*; 76, *129*; 109, *183*; 78, 132, *205*
Bhatt, B. D. 88, 112, *147*
Bird, A. F. 4, 6, 10, 11, 12, 14, 15, 16, 17, 26, 27, 33, 38, 49, 82, 114, 116, 118, 120, 121, 122, 123, 125, 126, 133, 136, 137, 161, 163, 166, *4*; 42, 43, 39, *70*
Blake, C. D. 87, 127, *151*
Bonner, T. P. 15, *32*; 20, *41*
Bossche, H. Van den 63, *107*; 95, 96, 142, *160*
Boucek, M. 85, *143*
Bovien, P. 159, *231*
Brand, T. von 7, 49, 50, 51, 52, 53, 56, 57, 58, 59, 60, 62, 63, 65, 70, 72, 73, 75, 76, 77, 78, 79, 80, 82, 83, 84, 85, 139, *8*
Braun-Schubert, G. 106, *176*
Bremner, K. C. 9, 47, 53, 114, 139, *12*
Brenner, S. 32, 134, 135, 136, *56*; 32, 136, 138, 161, 162, 163, 164, 165, 166, *59*; 139, 140, 141, 162, *212*
Briehl, R. W. 83, *137*
Bronner, R. 49, *76*
Bruce, R. G. 32, 34, *57*
Bryant, C. 62, 63, 65, 68, 69, 70, 72, *105*
Bryant, V. 88, *146*

Cain, G. D. 79, *131*
Cairns, E. J. 88, *148*
Castillo, J. Del 94, 95, 96, 133, 138, 139, 140, 141, 142, 144, *158*
Castillo, J. M. 65, 70, *114*
Castro, A. A. 107, *179*
Castro, C. E. 96, 107, *163*; 103, *172*
Castro, G. A. 59, 72, *98*
Chan, J. 10, *21*
Chance, B. 70, 86, *119*
Cheah, K. S. 73, *122*
Chitwood, B. G. 6, 7, 9, 32, 37, 38, 56, 116, 118, 120, 122, 125, 136, *10*
Chitwood, M. B. 6, 7, 9, 32, 37, 38, 56, 116, 118, 120, 122, 125, 136, *10*
Chvapil, M. 85, *143*
Clark, S. A. 12, 13, *27*; 116, 164, *191*
Clarke, A. J. 82, 123, *135*; 130, *201*
Cobb, N. A. 1, *1*
Colam, J. B. 31, 46, 50, *53*; 33. 38, 49, 52, 53, *60*; 46, *74*; 50, 52, *79*
Cole, R. J. 79, *132*
Cooper, A. F. 58, 64, *95*; 59, 64. 78, 80, 91, *99*; 90, *152*
Costello, L. C. 73, *123*
Cox, P. M. 82, 123, *135*
Crofton, H. D. 1, 7, 12, 29, 145, *2*; 7, 12, 27, 29, 56, 135, 140, 141, 145, 147, 156, 168, *9*; 36, 113, 136, 137, 141, *189*; 152, 175, *222*
Croll, N. A. 39, 77, 80, *65*; 162, 163, *233*; 167, 168, 174, *236*; 169, *237*; 169, 177, *238*; 172, 173, 176, *240*
Crompton, D. W. T. 70, 73, 89, *120*

Daems, W. T. 12, 26, *26*
Dankwarth, L. 106, *176*
Dart, P. J. 12, 13, *27*
Davey, K. G. 22, *43*; 22, *44*; 22, *46*
Dew, M. 139, 140, 141, 162, *212*
Doncaster, C. C. 35, 36, *62*; 38, 56, *63*; 39, 41, 177, *66*; 41, *68*: 125, *194*; 126, *196*
Dropkin, V. R. 31, 40, 42, 43, 51, *51*; 44, *72*
Dusembery, D. B. 170, *239*

Ehrlich, E. 85, *143*
Ellenby, C. 15, 82, 125, *31*; 104, *174*; 127, *197*
Epstein, H. F. 32, 134, 135, 136, *56*
Etges, F. J. 20, *41*

Evans, A. F. F. 128, 129, *200*; 167, 168, 174, *236*

Fairbairn, D. 51, *82*; 58, 75, 78, 80, *97*; 59, 65, 75, *100*; 60, 79, 102, *102*; 60, 72, 79, 122, 123, *103*; 60, 63, 77, 78, 92, 132, *109*; 64, 65, 70, 75, *110*; 79, *131*; 107, *179*; 109, *182*; 132, 138, *206*
Ferguson, J. D. 59, 72, *98*
Fernando, M. A. 62, 75, *106*; 83, *136*
Fodge, D. 108, *180*
Foor, W. E. 118, 120, *192*
Fredericks, W. 73, *123*
Fuchs, S. 14, *29*
Fujimoto, D. 14, 82, *30*; 15, 85, *33*

Gertz, K. H. 106, *176*
Gordon, R. 45, *73*; 53, 84, *87*
Govorova, S. V. 95, *161*
Gray, J. 148, 149, *219*; 147, *220*
Green, C. D. 9, 120, 170, 172, 175, *16*; 9, 120, *17*; 170, *193*
Greet, D. N. 9, 120, *17*
Griffin, F. D. 173, *242*
Grootaert, P. G. 16, 18, *36*
Gunter, B. L. 17, *39*
Guttekova, A. 1, *3*

Harpur, R. P. 54, 105, 108, *89*; 103, *173*
Harrington, W. F. 14, *29*
Harris, B. G. 108, *180*
Harris, J. E. 1, 7, 12, 29, 145, 147, *2*
Hechler, H. C. 44, *71*
Hess, R. J. 84, *141*
Hirumi, H. 134, *209*
Hobson, A. D. 102, 105, 106, 108, 113, *171*
Hope, W. D. 11, 12, *23*; 133, *207*
Hutchinson, G. W. 62, 75, *106*
Hutchinson, M. T. 158, *228*
Hutchinson, W. F. 65, *115*

Janssens, P. A. 110, *184*
Jenkins, W. R. 158, *228*
Jennings, J. B. 33, 38, 49, 52, 53, *60*
Johnson, P. W. 16, 18, 21, 27, *38*
Jones, F. G. W. 130, 159, *202*, 153, *223*; 153, 155, *224*
Jones, M. G. 130, 159, *202*
Jones, N. O. 16, *37*; 134, *209*

Kampfe, L. 17, *39*
Kan, S. P. 22, *43*
Kanaya, S. 14, 82, *30*
Kaplan, K. L. 93, *155*
Keilova, H. 56, *94*
Keith, R. K. 9, 47, 53, 114, 139, *12*
Kempton, A. 116, 164, *191*

Klewkowski, R. Z. 87, *145*
Klinger, J. 91, *153*; 176, *243*
Kolkofen, U. P. 47, 50, *75*
Körting, W. 64, 65, 70, 75, *110*
Kozek, W. J. 21, *42*
Krusberg, L. R. 58, 60, 77, *96*; 65, *114*; 75, *126*; 79, *132*
Kummel, G. 106, *176*
Kunz, P. 91, *153*

Langer, B. W. 75, *127*; 83, *140*
Lee, D. L. 4, 11, 12, 14, 15, 82, 92, 146, *5*; 4, 11, 12, 13, 16, 18, 27, 146, *6*; 9, 47, 52, 53, 110, 114, 139, *11*; 12, 13, 14, 33, 47, 52, 53, 139, *24*; 12, 19, 20, *25*; 50, 155, 177, *77*; 52, 83, 85, 92, 93, *83*; 52, 80, *85*; 80, *134*; 92, *154*; 101, *169*; 156, 157, *226*; *172*, *178*, *246*
Leštan, P. 51, *81*
Lippens, P. L. 10, *20*; 10, *22*; 16, 18, *36*
Lissman, H. W. 148, 149, *219*
Love, R. J. 9, 47, 53, 139, *13*

McElroy, F. D. 41, *69*
McLaren, D. 9, 161, 162, 163, 164, 165, 166, 167, *14*
McNeil, K. M. 65, *115*
Mapes, C. J. 23, *47*; 32, 36, 46, *54*
Marks, C. F. 103, 107, *172*
Matthews, B. E. 50, 155, *78*
Mayoh, H. 65, 75, 80, *116*
Menefie, M. G. 20, *41*
Mercer, E. K. 88, *148*
Miller, C. W. 84, *142*
Myers, R. F. 51, *80*; 105, *175*

Nelmes, A. J. 143, *213*
Newall, D. R. 83, 97, 98, 100, 102, 103, 105, 106, 107, 109, 111, 112, 113, *139*
Nicholas, W. L. 55, 59, 64, 77, 78, 79, 80, 82, 84, 85, 99, 118, 121, 122, *92*
Nicholls, F. 65, 72, *117*
Nicholls, P. 65, 72, *117*
Noble, E. R. 160, *232*
Noble, G. A. 160, *232*
Nolan, J. 9, 47, 53, 114, 139, *12*

Ogilvie, B. M. 9, 47, 53, 114, 139, *12*; 9, 47, 53, 139, *13*
Okazaki, T. 83, *137*
Orion, D. 49, *76*

Paltridge, R. W. 110, *184*
Pannikar, N. K. 102, *216*
Paplinska, E. 87, *145*
Parijs, O. F. J. Van 63, *107*
Parrott, D. M. 153, 155, *224*
Pasternak, J. 18, 21, *40*

Patourel, G. N. J. le 107, *177*; 107, 143, *214*
Perez-Gimenez, M. E. 83, *138*
Pitcher, R. S. 176, 177, *244*
Poinar, G. O. 84, *141*
Pollack, J. K. 109, *182*
Polyakova, O. I. 95, *161*
Poole, W. J. 76, *130*
Popkin, J. S. 54, 105, 108, *89*; 103, *173*
Pritchard, R. F. 96, *162*
Prokop, D. J. 15, 85, *33*
Prosser, C. L. 97, 103, 105, 109, 168, 169, 174, 175, *164*

Raski, D. J. 16, *37*; 134, *209*
Reed, E. M. 158, *227*
Reznik, G. K. 58, *86*
Riding I. 13, 53, 107, *28*
Riley, J. 52, *84*
Roberts, R. N. 84, *142*
Robinson, J. 159, 173, *230*
Rock, C. O. 55, *91*
Rogers, W. P. 16, 22, 126, 129, 131, 132, *34*; 16, 22, 23, 24, 26, *35*; 22, *45*; 23, 24, 46, 47, 52, 77, 80, 86, *48*; 110, 111, *186*
Roggen, D. R. 16, *37*; 32, 136, *55*; 34, *61*
Rohde, R. A. 88, 112, *147*
Rose, J. E. 93, *155*
Rosenbluth, J. 5, 134, 135, *7*; 133, 136, *208*; 135, *210*
Rothstein, M. 65, 75, 80, *116*; 65, 72, *117*; 79, *133*; 109, 110, *181*
Rothwell, T. L. W. 9, 47, 53, 114, 139, *12*; 9, 47, 53, 139, *13*
Rutherford, T. G. 53, *88*

Samoiloff, M. R. 18, 21, *40*
Sanborn, R. C. 107, *178*
Sanderson, B. C. 94, 95, 96, 142, 143, 144, *159*
Saz, H. J. 15, *32*; 63, 65, 67, 68, 69, 72, 74, 76, *108*; 72, 73, 142, *121*; 74, 95, 96, *124*
Schanbacher, L. M. 54, *90*; 55, *91*
Schindler, E. 70, 86, *119*
Schnitzerling, H. J. 9, 47, 53, 114, 139, *12*
Schoener, B. 70, 86, *119*
Seymour, M. K. 38, 56, *63*; 38, *64*; 39, 41, 177, *66*; 41, *67*; 41, *68*; 125, *194*; 145, *217*
Shepherd, A. M. 12, 13, *27*; 38, *64*; 32, 123, *135*; 116. 164, *191*; 126, *196*; 130, *201*
Shrimpton, D. H. 70, 73, 89, *120*
Silver, I. A. 70, 73, 89, *120*
Slocombe, J. O. D. 24, *49*

Smith, J. M. 162, 163, *233*, 167, 168, 174, *236*
Smith, L. 15, 82, 125, *31*
Smith, M. H. 52, 83, 85, 92, 93, *83*; 89, 90, 91, *150*; 92, *154*
Smith, W. J. 75, *127*
Smith, W. N. 73, *123*
Sommerville, R. I. 16, 22, 126, 129, 131, 132, *34*; 16, 22, 23, 24, 26, *35*; 22, *44*
Sproston, N. G. 102, *216*
Spurr, H. W. 143, *215*
Stauffer, H. 158, *229*
Stephenson, W. 100, 103, *168*; 102, 105, 106, 108, 113, *171*
Stolzy, L. H. 90, *152*
Streu, H. T. 158, *228*
Sulson, J. 139, 140, 141, 162, *212*

Talent, J. M. 108, *180*
Tarr, G. E. 60, 79, 124, *101*; 60, 79, 102, *102*
Theodorides, V. T. 75, *127*
Thienpont, D. 63, *107*
Thomas, R. J. 128, 129, *199*
Thomason, I. J. 96, 107, *163*; 103, 172
Thomson, N. 32, 136, 138, 161, 162, 163, 164, 165, 166, *59*
Thomson, W. W. 16, 18, 21, 27, *38*
Tomášek, V. 56, *94*
Triantophyllou, A. C. 10, 43, 122, *18*; 10, 43, 121, 122, *19*

Vaastra, W. J. 70, *118*
Van Gundy, S. D. 16, 18, 21, 27, *38*; 41, *69*; 58, 64, *95*; 59, 64, 72, 78, 80, 91, *99*; 90, *152*
Viglierchio, D. R. 97, *165*

Wallace, H. R. 27, 39, *50*; 99, 153, 154, *167*; 128, 150, 151, 153, 154, 159, *198*; 147, 150, 153, 154, 157, *218*; 150, *221*; 153, 156, *225*; 158, *227*
Wang, E. J. 72, 142, *121*
Ward, C. W. 58, 75, 78, 80, *97*; 59, 65, 75, *100*
Ward, K. H. 51, *82*
Ward, S. 32, 136, 138, 161, 162, 163, 164, 165, 166, *59*; 164, 170, 171, *234*
Warren, L. G. 74, *125*; 76, *130*
Wasilewska, L. 87, *145*
Waterson, R. H. 32, 134, 135, 136, *56*
Webster, J. M. 45, *73*; 53, 84, *87*; 53, *88*
Weinstein, P. P. 15, *32*; 44, 84, *93*; 102, 113, 114, *188*
White, J. G. 32, 136, 138, 161, 162, 163, 164, 165, 166, *59*
Wilson, P. A. G. 125, *195*

Winslow, R. D. 99, *166*
Wisse, E. 12, 26, *26*
Wittenberg, B. A. 83, *137*
Wittenberg, J. B. 83, *137*; 93, *156*
Wright, D. J. 83, 97, 98, 100, 102 103, 105, 106, 107, 109, 110, 111, 112, 113, *139*; 107, *177*; 110, *185*; 111, 112, *187*; 107, 143, *214*
Wright, K. A. 10, *21*; 11, 12, *23*; 31, 48, *52*; 32, 34, 138, *58*; 138, 163, *211*; 166, *235*

Zmoray, I. 1, *3*

Subject Index

Bold type indicates reference to an illustration or table.

Absorption 38, 45, 53–5
Acanthocheilonema, glycogen in 58
Acanthonchus, cuticle 11–12
ocelli 167
Acetylcholine 134, 138, 139, 140, 142, 143, **143**
nerve–muscle junction, effect on 139, 143, **143**
Acetylcholinesterase 52–3, 110, 114, 139, 142–3, 165
amphids, secretion from 165
anthelmintics, effect of 142–3
excretory system, glands of 52–3, 114, 139
nematicides, effect of 143
nervous system 139
secretion of 52–3, 110, 114, 139, 165
sub-ventral glands 52–3, 114, 139
Achromobacter nematophilus in *Neoaplectana* 45
Acrobeles, feeding 39
mouth parts 29
Actin 134
Actinolaimus, buccal cavity 30, **30**
enzymes in 48
feeding 44
mouthparts 30, **30**
Adenylate energy charge 130–2, **131**
Air swallowing 90–1
Alae 13
Alimentary canal, structure 28–38, **29, 30, 32, 33, 35, 36,** 37, 48, 53, 136
muscles of 6, 31–2, **33**, 34–5, 37, 38, 48, 55, 56, 135–6
γ-Aminobutyric acid (GABA) 140, **140**, 142, 143
Ammonia 83–4, 109–13, **111**
excretion of 109–13, **111**
Amphids 2, 3, 4, 15, 56–7, 136, **137**, 139, **161**, 164–5, **165**, 170–1
acetylcholinesterase in 139
anticoagulant in 56–7, 165
chemoreception, role in 164, 170–1
gland 165, **165**
moulting 4, 15
structure 164–5, **165**
Amylase 49, 50
Anabiosis *see* cryptobiosis
Ancylostoma, anticoagulant in 56–7, 165
buccal cavity 30, **30**
cuticle 146
excretory products 76
feeding 39, 46–7, 50
intestine 37
light response to 177
metabolism 83
oxygen ingestion 86
osmoregulation 102, **102**, 103, 113, 114
P/O ratio 74
skin penetration 155, 177
temperature, response to 88, 173, 177
touch, response to 177
Angiostrongylus growth 26
Anguina, cryptobiosis 103, 151, 159
desiccation survival 103, 151, 159
Angusticaecum, osmoregulation 101–2
Animals, location of 177–8
Anisakis, feeding 47
Anticoagulant 46, 47, 48, 56–7, 165
Antienzymes 56–7
Anthelmintics 83–6, 142–4
cholinesterase, effect on 142–3
fumarate reductase, effect on 95–6
glucose uptake, effect on 95–6
glycolysis, effect on 94
metabolism, effect on 95–6
neuro-muscular system, effect on 142–4
oxidative phosphorylation, effect on 95–6
resistance to 96
succinate formation, effect on 95–6
tricarboxylic acid cycle, effect on 95–6
Aphelenchoides, carbon dioxide, response to 176
cryptobiosis 159
defecation 38, 56
egg laying 125
feeding 44, 151
glycogen 58
locomotion 151, 176
metabolism 65, 84
movement 176
osmoregulation 99
oxygen and development 89
permeability 107
pharyngeal pumping 34–5
sugars 60
Aphelenchus, ethanol formation 64
feeding **40**, 42
glycogen 58, **59**
glycogen consumption 59
hormones, effect on 22

lipid **77**
lipid consumption 59, 77–8, 80
metabolism 64, 80
moulting 21
oxygen debt 64
oxygen, lack of 90
permeability 103
pharynx 32
quiescence 64, 91
Aplectana, feeding **36,** 39
Aporcellaimellus, cuticle formation 17
moulting 16, 17, 21
Ascaridia, anthelmintics, effect on 95–6
ascarosides 60, 79
excretion 110, **111**
feeding 45–6
glycogen 58, **59**
glycogen consumption 58
hatching 90
metabolism 65, 84
nitrogenous excretion **111**
tricarboxylic acid cycle 65
secretion in 53
urease 110
Ascaris, acetylcholine 138–9
acetylcholine, response to 139, **143,** 143
adenylate energy charge of eggs 132
anthelmintics, effect on 95–6, **140,** 142–4, **143**
antienzymes 56
ascarosides 60, 124
buccal cavity 30
catecholamines 140
chitinase 51
collagen 14, **14,** 85
cuticle 12, **12,** 15, 145–6, **146**
composition of 14–15, **14**
enzymes in 15
growth 12, 26–7, **26**
structure 12, **12,** 145–6, **146**
cytochromes 72–3
digestion 51, 52, 53
egg 60, 73, 75, 78, **78,** 83, **123,** 124, 132
electron transport 73
energy charge **78,** 132
glyoxylate cycle 75
metabolism 78, **78**
permeability 124
egg production 83
egg shell, composition 60
electron transport 72–3
excretion 69, 76, 109–13, **111,** 112
excretory products **69,** 76, 109–13, **111,** 112
excretory system **6,** 9
feeding **36,** 45–6, 47
fumarate reductase 67–70
gluconeogenesis 75
glucose uptake 53–4, 108
glycogen, characteristics 59
glycogen consumption 58, 59
glycogen content 58, **59**
glycolysis 62, 65
glyoxylate cycle 75, 80
growth 3, 25–7, **26**
haemoglobin 82–3, 85, 86, 91–3, **92**
haemoglobin, digestion of 52
hatching 90, 125, 126, 128, 132
hydrostatic pressure 28, 145
invertase 51
ionic regulation 105–7, 113, 114
lactic acid 67
lipase 52
lipid consumption 78, **78,** 79
lipid content **77,** 78
lipid, in eggs **78,** 78, 80
locomotion 145–6, 155
maltose 51
metabolism 62–3, 65–70, **68, 69,** 72–6, 78, **78,** 79–80, 83–5, 109–10
metabolic rate 87
moulting 17, 21
mouth parts 28, 45–6
movement in intestine 155
muscle 76, **134,** 134, **135,** 138
nervous system **137,** 138
nitrogenous excretion 109–13, **111**
nutrient uptake 53–5, **54,** 108
osmoregulation 103
oxygen affinity **92**
oxygen consumption 90
oxygen for development 89
oxygen ingestion 86
Pasteur effect 63
pentose–phosphate pathway 75
peptidases 52
permeability 107
pharynx 31–2, **32,** 36, **36**
phosphate/malate ratio 74
P/O ratio 74
phosphogens, lack of 76
pseudocoelomic fluid 7
pseudocoelomocytes 7
secretion in 53
spermatozoa 120
sugars 60, 62
trehalase 51
tricarboxylic acid cycle 65–70, 73
urease 110
volume regulation 102, 103
ascarosides 60, 78–9, 122, 124, **124,** 132

Aspiculuris, hatching 132
 5-hydroxytryptamine 118–19
 intestine 37
 osmoregulation 102
 sex attraction 119
 spermatozoa **119**

Bacillary bands 10
Behaviour 39–41, 161, 168–78
 chemicals, response to 39–41, 169–72
 electrical field, response to 174–5
 escape 172
 feeding 39–41
 gravity, response to 175
 light, response to 173–4
 moisture, response to 175
 temperature, response to 172–3
 touch, response to 172
 whole patterns of 175–8
Bile, in exsheathment 23
Biological control 45
Body cavity *see* pseudocoelom
Body size, and respiration 86–7
Bradynema, cuticle, lack of 13
 feeding 45
 nutrient uptake 45, 53, 107
Brugia, lactic acid 63
Buccal cavity 1, **2**, 7, 28–31, **29**, **30**, 46, 47
Bunostomum, feeding 46
 mouthparts 29, **29**
Bursa, copulatory **2**, **9**, 116, **118**, 120, 136, 164
 sensilla 164

Caenorhabditis, amino-acid synthesis 84–5
 amphids 164, **165**
 catecholamines 140–1
 chemical transmitters 140–1
 chemoreception 164–6, 170–1, **171**
 ethanol production 64
 glycogen content 58, **59**
 glyoxylate cycle 65, 80
 growth *in vitro* 55
 hermaphrodites 121
 hormones, effect of 22
 lipid consumption 77–8
 lipid content 77
 metabolism 64, 65, 79, 80, 84–5
 muscle 32, 135
 nervous system 138, 162–6
 nitrogen catabolism 109
 oxygen debt 64
 oxygen, lack of 90
 pharynx 32
 quicscence 91
 sense organs 162–6, **162**, **165**, **166**

Camallanus, haemoglobin 93
 sex attraction 119
Capillaria, bacillary band cells **8**, 10
 cuticle 146
 feeding 47–8
 gland cells **8**, 10, 34
 pharynx 32, 34
 stichocytes 34
 stylet 31, 48
Carbohydrate, content 58–60, **59**
 distribution 58–60
 metabolism 58–76, **61**, **66**, **68**, **69**, **71**, **74**, 78, **78**
 metabolism in egg 78, **78**
Carbon dioxide, as attractant 172
 excretion 112
 exsheathment, role in 23–4
 hatching, role in 132
 oxygen consumption, effect on 112
Carbonic acid, exsheathment, role in 23–4
 hatching, role in 132
Carbonic anhydrase 52, 112
Carnivorous, feeding 28, 30, **30**, 43–4
Catecholamines 140–1, 142, 162
 sensilla 162
Caudalids 137
Cell constancy 27
Cellulase 49, 51
Cephalids 137
Cephalobus, pharynx **32**
Chabertia, buccal cavity **30**
 feeding 46
Chaetosoma, movement 158
Chemoreception 164–7, **165**, **166**, 169–72, **170**, **171**, 176, 177
 feeding, role in 39–41, 169
Chemoreceptors 164–7, **165**, **166**, 176, 177
Chitin 13, 60, 82, 122, 123, **123**, **124**, 126
 synthesis 60
Chitinase 49, 51, 126
 hatching, role in 126
Chromadorina, glands, hypodermal 10
 sensilla 163
Circadian rhythm, microfilariae 160
Cloaca 1, **2**, **9**, 38, 116, **116**
Collagen 1, 12, 14, **14**, 19, 20, 82, 85
 Ascaris 14, **14**, 85
 composition **14**, 85
 distribution 82
 egg shell 123
 molar mass 85
Contracaecum, feeding 50
Copulation 116, 118, 119, 120, 137, 164
 control of 137
 sensilla, role of 164
 spicules, role of 116

Copulatory bursa 2, 9, 116, **118**, 120, 136, 164
sensilla 164
Cooperia, metabolism 82
Cosmocerca, feeding 45–6, 50
Crawling 149–52, **150, 151**
Criconemoides, feeding 158
movement 158
Cryptobiosis 1, 91, 103–4, 151, 152, 159
dispersion, role in 159
Cuticle 1, **3**, 3–4, **4**, 5, 11–27, **11**, **12**, 47, 82, 92, 107, 116, 124–5, 145–6, **146**, 158
attachment to hypodermis 5, **5**, 19, 20
collagen 1, 12, 14, **14**, 19, 20, 82, 85
composition 14–15, **14**, 82, 85
dissolution of 16–17, **17**, 21
enzymes 15, 27
feeding, role in 45, 47, 53
formation 16–21, **17**, **19**, **20**
growth 12, 26–7, **26**
haemoglobin 15, 92
juvenile 11–14, **11**
locomotion, role in 145–6, **146**, 158
permeability 107
pores 13
ridges **4**, 13, 14, 47, 157
spicule 116
structure 3–4, **4**, 11–14, **11**, **12**, 18
tanning 14–15, 124–5
Cuticlin **14**, **14**, 82
Cyathostoma, feeding 46
Cytochromes **71**, 72–3

Dauer juvenile 2, 15–16, 22, 25, 159
ecdysis 15–16, 22
exsheathment 15–16, 22
Defecation 38, 41, 56, 135, 137, 147
Deirids 162, 163
Delodanus, feeding 44–5
Deontostoma, ocelli 167, **167**
Desiccation 1, 103–4, 151, 152, 159
Desmoscolex, movement 158
Diapause 129
Dictyocaulus, dispersal 159, 173
exsheathment 23
fumarate reductase pathway 69
metabolism 69, 70, 84
light, response to 173
Digestion 31, 37–8, 39, 46, 48–53
carbohydrates 48–51
enzymes 48–53
lipids 48, 50, 52
plant cells 31
prey 31
proteins 48, 50, 51–2
Dioctophyme, feeding 47
lipid content **77**
Dipetalonema, spermatozoa 120
spicules 116, 164
Diplogaster, enzymes 48
feeding 48
Dirofilaria, amino-acid uptake 84
anthelmintics, effect on 95
feeding 48
glycogen content **59**
lactic acid 63
metabolism 63
Pasteur effect 63
Dispersal 2, 158–60
earthworms, role of 160
insects, role of 159, 160
passive 159
secondary hosts, role of 160
vectors, role of 160
Ditylenchus, amino-acid synthesis 85
copulation 120
cryptobiosis 103–4, 159
cuticle 146
desiccation 103–4, 159
dispersal 159
enzymes 49, 50–1
excretion **111**
feeding 39, **40**, 41, 42, 49
fumarate reductase pathway 70
lipid content **77**
locomotion 153
metabolism 65, 70
oxygen consumption 88, 112
oxygen and development 89
pharyngeal pumping 34–5
sugars **60**
temperature, response to 173
tricarboxylic acid cycle 65
Dopamine 140–1
Dorylaimus, buccal cavity 30, **30**
feeding **30**, 44
pharynx **32**
Dracunculus, growth 3
excretion 112
lactic acid 112

Ecdysis 15–25, **24**
see also exsheathment
Egg 2, **2**, 9, 78–9, 78, 80, 104, **121**, 122 **123**, 124, **124**, **126**, 30–2, **131**, 155–6
Ascaris 78, **78**, **123**, 132
ascarosides 60, 78–9, 122, 124, **124**, 132
desiccation 104
development 122
emergence from 125–7
energy charge 130–2, **131**
hatching 125–33
Heterodera **124**

Egg (cont.)
 laying 9, 125
 Mermis 123
 metabolism 73, 78, **78**, 80
 movement in 155–6
 oxyuroid 123, **123**
 permeability 124
 structure 9, 121, 122–5, **123**, **124**
 Trichuris 123
 tylenchid **123**
Egg shell, chitin 9, 123, **123**, **124**, 127
 collagen 123
 composition 9, 60, 82
 formation 121, 122–5, **124**
 permeability 124
 structure 9, 121, 122–5, **123**, **124**
Electron transport 60, 70–3, **71**
Embden-Meyerhof pathway *see* glycolysis
Energy charge 130–2, **131**
Enoplus, feeding 93, **94**
 haemoglobin 91, **92**, 93, **94**
 ionic regulation 105, **106**, 107
 metabolic rate 90
 muscles 32, 135
 osmoregulation 97–8, **98**, 102
 oxygen affinity **92**
 oxygen consumption 88, 90
 oxygen and feeding 93, **94**
 permeability 103
 pharynx 32, **32**
 quiescence 91
 thigmokinesis 172
 volume regulation 97–8, **98**, 103
Ensheathed juveniles 2, 22–5, **24**, **25**, 104
Enterobius, glycogen consumption 59
 anthelmintics, effect on 143
Entomophagous, feeding 44–5
Enzymes, hatching 125–6
 host penetration 155
 intestine 37–8, 50–2
 pharynx 31, 33, 34, 47, 48–50
 secretion **40**, 46, 47, 48, 49, 50, 52–3, 110, 114
 sub-ventral glands 47, 52–3
Epicuticle 11, **11**, 16, 18, **19**
Ethanol, formation **61**, **64**
Eudorylaimus, flagella 1, 147
Eustrongylides, glycogen content 58, **59**
 haemoglobin 93
 lipid content 77
Excretion 76, 109–15, **111**
 ammonia 109–13, **111**
 dyes 114
 lactic acid 63
 nitrogenous 109–13, **111**
 urea 110, **111**, 113
Excretory canal **3**, 4, 8–9, 114, **117**
Excretory gland **2**, 8–9, 21–2, 47, 113–15
 see also sub-ventral gland
 enzymes 9, 12–22, 47, 139
Excretory products 63–4, 67–9, **68**, **69**, 76, 83–4, 109–13, **111**
 acetylcholinesterase in 52–3, 114, 139
 excretion 113–15
 extracorporeal digestion, role in 114
 functions of 113–15, 125
 gelatinous matrix, secretion of 114
 glands **2**, 21, 52–3, 113–15, 139
 ionic regulation 113–15
 moulting, role in 21–2, 24, 114
 osmoregulation 102, **102**, 106, 113, 115
 structure **2**, **3**, 4, **4**, **6**, 8–9, 113, 115, **117**
 types of **6**, 8–9
Exsheathing fluid 21–2, 23, 24
Exsheathment 21–5, **23**, **24**
 carbon dioxide 23–4, **23**
 enzymes 21–4
 pH 23–4, **23**
 receptor 24
 stimulus 15–16, 22–5, 23
Extracorporeal digestion **40**, 41, 47, 48, 50, 51, 52, 114

Fat *see* lipid
Feeding **frontispiece**, 1, 22, 28, 30–1, **30**, **33**, 34–6, **35**, **36**, 38–48, **40**, 48–53, 158, 177
 animal-parasitic **29**, 30, 31, 33, **36**, 45–8, 50, 52–3
 carnivorous 28, 30, **30**, 43–4, 51
 chemoreceptor 169
 entomophagous 44–5, 48
 length, changes during 135
 microbivorous **29**, 30, **30**, **33**, **35**, **36**, 38–9, 43, 45, 46–7
 mycophagous 39–43, **40**
 oxygen consumption 88
 phytophagous **frontispiece**, 30–1, **30** 34–5, 39–43, **40**, 49, 50–1
 saprophagous 28, **35**, **36**, 38–9
 searching behaviour 168
 sense organs 171
Fertilisation 121–2
Filariae, dispersal 160
 egg shell 124
 feeding 48
Flagella, intestinal 1, 147
Fumarate reductase pathway 65–70, **68**, 95–6
 anthelmintics, effect on 95–6

Giant cells (transfer cells) 31, **40**, 41, 43, 49

Gland cells, amphid 165, **165**
caudal 10, 120
excretory **2**, 21, 52–3, 113–15, 139
hypodermal **8**, 10
pharyngeal 7, 31, 32–3, **32**, **33**, 39, **40**, 41, 46, 47, 48, 49, 50
rectal **2**, 10
Gluconeogenesis 62, 75
Glucose uptake, anthelmintics, effect on 95
Glycogen, content 58–9, **59**
chain length 59
heat of combustion 59
molar mass 59
Glycogenesis 75
Glycolysis 60, **61**, 62, 95
anthelmintics, effect on 95
Glycosides 60, 78–9, 124
Glyoxalate cycle 65, 75, 80, **81**
Growth 3, 25–7, **25**, **26**, 55
Gubernacalum 3
Gustatory sense organ 166–7

Haemoglobin 4, 7, 50, 51, 82–3, 85, 86, 91–3, **92**, **94**, 167, 174
composition 85
cuticular 92
digestion of 50, 51, 85, 91
distribution 82–3
functions 83, 91–3, **94**, 167, 174
molar mass 83, 85, 92
oxygen affinity 83, 91–3, **92**
oxygen diffusion 86
oxygen store 93, **94**
pharynx 93, **94**
photoreception 167, 174
Haemonchus, anthelmintics, effect on 95–6
anthelmintic resistance 96
cryptobiosis 104, **104**, 151
desiccation survival 104, **104**, 151
excretion 110
exsheathment 16, 22–4, **23**, **24**, **25**
feeding 47
haemoglobin **92**, 93
hormones 22
leucine aminopeptidase 16, 24
metabolism 62, 63, 95–6, 109–10
nitrogen catabolism 109–10
oxygen affinity **92**
oxygen consumption 90
Pasteur effect 63
Hammerschmidtiella, osmoregulation 101, **101**
volume regulation 101, **101**
Hatching 49, 89–90, 125–32, **126**, **127**, **131**
carbon dioxide, effect of 132
chemicals, effect of 126, 129–32
energy charge 130–2, **131**
environmental effects 127–8, **127**
enzymes, role of 49, 125–6, 132
host specificity 129–32
metabolism 130–2, **131**
moisture, effect of 127–8
oxygen, effect of 89–90, 127–8, **127**, 130–2, **131**
pH, effect of 132
stimuli 125–32
stylet, role of 126, **126**
suppression of 128
reducing agents, effect of 132
temperature, effect of 128–9
Head **2**, 6, 7, 28–31, **29**, 30, 41, 161–3, **161**, **162**, **165**, **166**
sensilla **2**, 161–3, **161**, **162**, **165**, **166**
Hemicycliophora, feeding 42
moulting 16, 18
Hemizonion 137
Hemizonoid 137
Hermaphrodites 2, 121, 122
Heterakis, excretory canal **117**
egg shell formation **124**
feeding 45–6
fumarate reductase pathway 69
glycogen content **59**
haemoglobin 93
intestinal muscle 56
intestine 37, 56
mouth parts 28, **29**
pharynx **33**
sensilla **117**
spermatozoa **119**
spicules 116, **117**
spicules, innervation 164
Heterodera, behaviour, host finding 177
behaviour, sex 9, 119–20, 169–70, **170**, 172
cuticle 13, 15, 124–5, 146
cyst 13, 15, 124–5, 128–31, **131**
diapause 129
desiccation survival 104
dispersal 159
egg, drying of 159
egg, gelatinous mass 125
egg shell 82, 123, 126
energy charge 130–1, **131**
feeding 39, **40**, 41, 43
hatching 126, **126**, 127–31, **131**
host, penetration of 41–2
ionic regulation 106, 107
locomotion **151**, 153
moulting 17
osmoregulation 100–1, **100**
polyphenol oxidase 15, 124–5
sex attraction 9, 169–70, **170**, 172

Heterodera, (cont.)
spicules, function of 164
spicules, innervation 164
tanning 15, 124–5
volume regulation 100–1, **100**
Hexatylus, feeding 41
Hirschmaniella, cuticle 21, 146
moulting 21
Hormones 22
Host, location of 39–41, 169, 176–8
penetration of 39–41, **40,** 176–8
Hyaluronidase 49, 51
Hydrostatic pressure 34, 44, 145–7
locomotion 145
loss of 44
pharynx 34
physiology, effect on 145–7
pseudocoelom 145
structure, effect on 7, 28, 34–5, 38, 145–7
Hydrostatic skeleton 7, 28, 145–7
5-Hydroxytryptamine 118–19
Hypodermis 1, **3,** 4–5, **5,** 13, 15–20, **17, 19, 20,** 27, 107
moulting, role in 15–20, **17, 19, 20**
permeability, control of 107

Ingestion 31, 34–6, **35, 36,** 39–48, **40**
Intermediary metabolism 60–76, **61, 66, 68, 69, 71, 74, 81**
Intestine 1, **2, 3, 4,** 7–8, 28, 36–8, **37,** 50–5, **54,** 56, 82, 89–91, 103, 107, 108, 113, 147
absorption 38, 53–5
enzymes 37–8, 50–2
excretion, role in 113
flagella 1, 147
fluid transport 54, **54,** 108
muscle 37, 55, 56, 147
nutrient uptake 38, 53–5
osmoregulation, role in 103
oxygen in 89–91
permeability 107
secretion by 8, 38, 53
valves 8, 38, 56
Invertase 49, 51
Ionic regulation 88, 104–8, 113, 114, 115

Juvenile, ensheathed 2, 22–5, **24**
hormone 22

Kinesis 168, **168,** 170
Krebs' cycle *see* tricarboxylic acid cycle

Lactic acid **61,** 63–4, 67, 69, 112
Leidynema, ascarosides 60
egg **123**
feeding 45
peptidases 52
pharynx **32**
Leucine aminopeptidase 16, 21–2, 24, 52
ecdysis, role in 16, 21–2, 24
exsheathment 16, 24
moulting 16, 21–2, 24
Light, response to 167–8, 173, 174
Lipases 50, 52
Lipid, content 76–9, 77
carbohydrate, conversion to 78, **78** 80
metabolism 76–83, **78**
metabolism, in egg 78, **78**
Litomosoides, anthelmintics, effect of 95
electron transport 72
glycogen content **59**
lactic acid 63, 112
Pasteur effect 63
Locomotion 145, 160
animals, within 155, **156**
behaviour 168, **168**
deep fluids 147–9, **149,** 151, **151, 152**
egg, within 155–6
nervous system 137
oxygen 90
soil, through 152–5, **154**
water films 149–52, **150, 151**
Longidorus, feeding 39, 42
movement 153
muscle 134
pharyngeal glands 33
stylet, gustatory role 166–7

Maltose 50, 51
Mechanoreceptors 161–4, **161, 162, 163, 166, 167**
Meloidogyne, amino-acid synthesis 85
enzymes in 49
feeding **40,** 41, 43
gelatinous egg mass 125
growth **25,** 26
hatching 126–8, **127**
intestine 37
lipid 77–8, 77
locomotion 150
metabolism 77–8, 82
moulting 16–17, **17,** 21, **25,** 26
oxygen, effect on hatching 89
oxygen, lack of 90
parthenogenesis 121, 122
rectal glands 10, 38, 125
sensilla 163
sex determination 9–10
sex reversal 10, 43
Mermis, alimentary tract 37, 38, 45, 53
cuticle 13

egg **123**
haemoglobin 91, 167
light, response to 167, 174
locomotion 157
metabolism 82, 84
nutrient uptake 53, 84
sex ratio 9
Mermithids, alimentary system 45
feeding 45
see also Mermis
Mesodiplogaster, air swallowing 90–1
Metabolic rate 87–8, **87**, 112
body size, effect of 112
temperature, effect of 88
Metabolism 58–96
anthelmintics, effect on 95–6
nematicides, effect on 93–6
Microfilariae, dispersal 160
diurnal rhythm 160
feeding 45
Moisture, effect on hatching 127–8
Moisture characteristic, locomotion 153–5, **154**
Monhystera, swimming 148
Mononchus, buccal cavity 30, **30**
feeding 44
pharynx **32**
proteases 51
Morphology, general 1–10, **2, 3, 4, 5, 6, 29, 30, 117**
Moulting 2, 4, 15–25, **17, 19, 20**
dissolution of cuticle 16–17, **17**, 21
excretory system, role in 21, 24, 114
leucine aminopeptidase 16, 21–2
neurosecretion 21–2
stimulus 15–16
Movement 145–60, **148, 149, 150, 151, 152, 156, 157**, 168–78, **168, 176**
animals, within 47, 155, **156**
backwards 156
deep fluids 147–9, **149**, 151, **151**, 152
egg, within 126, **126**, 155–6
feeding 39–42, 47, 156
helical 156–8, **157**
jumping 158
soils, through **148**, 152–5
unusual forms 158
water films, in 149–52, **150, 151**
Muscle 1, **3, 4, 5**, 5–6, 9, 18, 20, 27, 31–2, **33**, 34–5, 37, 38, 39, 56, 76, 133–6, **133, 134, 135, 137**, 139–40, 145–7, 156–8, **157**
actin in 5, 134
alimentary tract 6, 31–2, **33**, 34–5, 37, 38, 39, 56, 136, **137**, 147
anthelmintics, effect on 95–6, 134–5, 142–4
contraction of 139–40, 145, 147
cuticle, attachment to 5, **5**, 6, 20
cytoplasmic bridges 6, 27
innervation **3**, 5, 133, **134**, 137
locomotion 145–7, 156–68, **157**
moulting 18, 20
myosin 5, 134
paramyosin 134–5
reproductive system 6, 9, **116**, 118, 121, 125, 136
resting potential 139
sphincter 38, 56, 136, **137**
structure **3**, 5–6, **5**, 133–6, **133, 134, 135**
Mycophagous, feeding 34, 39–42, **40**
Myosin 5, 134

Naccobus, feeding 43
Necator, acetylcholinesterase 139, 165
feeding 46
skin penetration 177
temperature, response to 88, 173, 177
Nematicides **94, 96**, 107, 142–3
cholinesterases, effect on 143
permeability, effect on 107
Nematodirus, diapause in 129
excretion 109, 110, **111**
exsheathment 23
feeding 47
haemoglobin **92**, 93
hatching 128, 129
movement 155
oxygen affinity **92**
Nematospiroides, copulation 120
metabolic rate 87–8
moulting 20
spermatozoa 120
Neoaplectana, Achromobacter in 45
bacterium, association with 45
feeding 45
growth *in vitro* 55
movement of 158
Nervous system **3**, 6–7, 133, 134, 136–44, **137**
catecholamines in 140
cholinergic 138
co-ordination in 137
nerves, inhibitory 141, 143
nerve–muscle junction 133, **134**, 138, 143
nerve ring **3**, 6, 136, 137, **137**, 138
nerves, motor 137, 140, 141
pharyngeal 7, 31, 32, **33**, 138
sensory 137, 140
structure **2, 3**, 6–7, 136–8, **134–137**
transmission along 138–42
Neurohumors 134, 138, 139, 140, **140**

Neuro-muscular physiology 133–44, **134**, **143**
Neurosecretion 21–2
Nippostrongylus, acetylcholinesterase 1, 2–3, 114, 139
behaviour 177
cuticle **4**, 12, 14, 18, 47, 146
cytochrome 72
electron transport 72
enzymes, digestive 51
enzymes, secretion 47, 52–3, 110, 114
excretion 112
extracorporeal digestion 52
feeding 39, 47
glands, pharyngeal 33
glands, sub-ventral (excretory) **4**, 8, 52–3, 114
glycogen content 58, **59**
haemoglobin 15, **92**, 93
immunity, effect on 79–80
intestine 37, **37**
lipid content 77, 79
locomotion 14, 155, **156**, 157
metabolism 63–4, 65, **69**, 72, 79–80, 82
moulting 15, 16, 18–20, **19**, **20**
osmoregulation 102, **102**, 103, 113, 114
oxygen affinity **92**
oxygen consumption 90
oxygen, lack of 90
oxygen uptake 88
Pasteur effect 63
quiescence 91
sensilla 163
sex attraction 172
skin penetration 155, 177
spermatozoa **119**
spicules, innervation 166
sugars 60
temperature, response to 88, 173, 177
thigmokinesis 172, 177–8
Nitrogen, catabolism 109–12, **111**
excretion **111**, 112–13
Nucleosides 75–6
Nucleotides 75–6
Nurse cells **40**, 43
Nutrient, requirements 55
uptake 53–5, **54**, 84, 107, 108
Nygolaimus, feeding 44

Obeliscoides, haemoglobin 83
Ocelli 167–8, **167**, 174
Oesophagostamum, acetylcholinesterase 52–3
excretory system **6**
Oesophagus *see* pharynx
Onchocerca, feeding 48
Oogenesis 120–2
Osmoregulation 88, 97–104, **100**, **101**, **102**, **104**, 113–5
animal-parasitic nematodes 101–2, **101**
excretory system, role in 102, **102**, 106
fresh-water nematodes 99–100
marine nematodes 97–9
oxygen consumption 88
plant-parasitic nematodes 100–1, **100**
soil nematodes 99–100
Ostertagia, hatching 128
Ovary **2**, **3**, 9, 82, 108, 121, **121**
nutrient uptake 108
Ovijector 121, **121**, 125
Ovoviviparus 124
Oxidative decarboxylation *see* TCA cycle
Oxidative phosphorylation *see also* electron transport 60, 70–3, **71**, **96**
anthelmintics, effect on 96
Oxygen, availability 88–91
body size 86
consumption 64, 78, **78**, 87, **87**, 90
debt 63–4
demand 86–8, **87**
diffusion 85–6, 93
facilitated 86, 93
feeding activity, effect on 88
hatching, effect on 89–90, 127–8, **127**, 130–2, **131**
haemoglobin 91–3
ingestion 86
intestine 70, 73, 89–91
ionic regulation 88
locomotion, effect on 64, 88, 90, 91
requirements 64, 65, 90, 91
soils and muds 89–91
Oxyuris, cuticle 12, 146

Panagrellus, cuticle 21, 146
desiccation survival 104
egg shell 124
excretion **111**
feeding **36**, 39
gluconeogenesis 75
glycogen 59, **59**
glyoxylate cycle 75–80
hormones, effect on 22
ionic regulation 105
lipid 59, 77–8, **77**
metabolism 65, 75–80, 87, 109
moulting 21
Panagrolaimus, metabolic rate 87–8
osmoregulation 99
sex attraction 119, 172

Paralysant 31, 44, 48
Paramyosin 134, 135
Parascaris, ascarosides 60, 79
egg shell composition 60
hatching 125
Parasites of animals, feeding 28, 33, 34, 45–8
Paratylenchus, moulting stimulus 15
Parthenogenesis 2, 121, 122
Passalurus, ascarosides 60, 79
Pasteur effect 63
Pectinase 49, 51
Pelodera, feeding 88
oxygen consumption 88
sex attraction 172
Penetration of host **40,** 41–2, 49, 50, 155
Pentose–phosphate pathway 74–5, **74**
Peptidases 50, 52
Permeability, cuticle 107
nematicides 107
solutes 107–8
water 103, **104**
Peroxidase 49
pH, exsheathment, role in 23–4, **23**
hatching, role in 132
Pharmacology 93–6, 139, 142–4
Pharynx 1, **2, 3,** 4, 7, 15, 28, 31–6, **32, 33, 35, 36,** 39, **40,** 41, 46–50, 93, **94,** 114, 126, 135, 138, 167, **167**
enzymes 31, 33, 34, 46, 47, 48–50
functioning of 31–6, **35, 36,** 39, 41, 42, 45, 46, 47, 48, 147
glands 7, 31, 32–4, **32, 33,** 41, 47, 48, 49, 50, 114, 126
glands, hatching, role in 126
glands, innervation 138
haemoglobin 93, **94**
innervation 32, **33,** 138
moulting 15
muscles 32, **33,** 34, 35, 135
ocelli 167, **167**
pumping 34–6, 41, 42, 46
secretion from 31–6, **32, 33, 40,** 41, 44, 46, 47–50, 114
Phasmids 3, 4, 15, **116,** 165–6
structure 165–6
Phocanema (=*Terranova*), hormones, effect on 22
moulting 21–2
neurophysiology 141
neurosecretion 21–2
temperature response to 173
Phosphagens 76
Photoreceptors 167–8, 174
Phytophagous, feeding **frontispiece,** 28, 31, 34–5, 39–43, **40**
Pilobolus, dispersal of *Dictyocaulus* 159, 173
Plant-parasitic, feeding **frontispiece,** 28, 31, 34–5, 39–43, **40**
Plants, location of 39–41, 176–7
P/O ratio 73–4
Polyphenol oxidase 124–5
Polyphenol quinone tanning 82
Pontonema, digestion 39, 49, 52
enzymes 52
feeding 39, 49
intestine, absorption 38
intestine, secretion 38
pharyngeal glands 33
secretion in 38, 53
Porrocaecum *see also Phocanema*
glycogen content **59**
lipid content **77**
spermatozoa 120
Pratylenchus, feeding 42
lipid content **77**
Predators *see* carnivores
Proctodaeum 28, 38
Proprio-receptor 141, 177
Proteases 48, 49, 50, 126
hatching 126
Protein, content 82
distribution 80
metabolism 80–5
Pseudocoelom 1, **2, 3,** 7, 28, 145
Pseudocoelomocyte **2,** 7
PK/PEPCK ratio 67–9, **69**

Quiescence 64–91

Radopholus, dispersal 159
feeding 39, 42
Rectal glands **2,** 10, 38, 125
Reproductive physiology 116–25
Reproductive system, structure 2, **2, 3,** 9–10, 116–9, 120–2, **116, 117, 118, 119, 121,** 136
Respiratory physiology 85–93
Response to, carbon dioxide 176
chemicals 169–172, **170, 171**
electrical field 174–5
external stimuli 168–78
gravity 175
heat 172–3
light 173
touch 172, 176, 177
Rhabdias, digestion 50
enzymes 52
excretory system **6**
Rhabditis, buccal cavity 30, **30**
cuticle 12
excretory system **6**
exsheathment 15–16

Rhabditis (cont.)
  feeding **35**, 38–9
  glyoxalate cycle 80
  osmoregulation 99–100, 103
  pharynx **32**
Root diffusates, hatching 129–31
Roots, attraction to 176–7, **176**
Rotylenchus, gelatinous egg mass 125
  feeding **40**, 42

Saliva, digestion 34, 41, 44, 46–50
  enzymes 46–50 *see also* pharynx
  hatching, role in 34, 126
  host penetration 34, **40**, 41–2, 49
Saprophagous, feeding 28, **35**, **36**, 38–9
Secretion, enzymes **40**, 46–50, 52–3, 110, 114
Seinura, feeding 44
Sense organs 1, 2, 3, 6, 28, 41, **117**, 161–8, 161, 162, 163, 165, **166**, 167
Sensilla 2, **2**, 116, **116**, **117**, 136, 137, **137**, 140, 161–7, **161**, **162**, **163**, **165**, **166**
  amphids **2**, **161**, 164–5, **165**
  cephalic 136, **137**, 140, **161**, **162**, 166
  chemoreceptors 164–7, **165**, **166**
  chordotonal 163
  copulation, role in 164
  copulatory bursa 163–4
  definition 161
  deirids 140, 162, 163
  genital 116, **116**, **117**, 163–4, 166
  gustatory 166–7
  innervation 136, 137, **137**
  labial 136, **137**, **161**, 162, 166, **166**, 177
  mechanoreceptors 161–4, **161**, **162**, **163**, 166, **166**
  spicule 116, 164
  structure 161–7, **161**, **162**, **163**, **165**, **166**
Sensory papilla *see* sensilla
Sex attraction 9, 10, 119–20, 169–70, **170**, 172
Sex determination 9–10, 43
  ratio 9
  reversal 43
Sinusoidal movement 147–56, **148**, **149**, **151**, **152**, 156
Skin penetration 49–50, 155, 173, 177
Soil, matric potential 99
  pore spaces 99
  water content 99
Spermatogenesis 116–9
Spermatozoa 2, 2, 9, 82, 116, 118, **119**, 120, 122
  fertilisation 122
  movement 118
  structure 118, **119**
  transfer of 116
Sphaerularia, nutrient uptake 53
Spicule 2, 2, 9. 38, 39, 82, 116–20, **116**, **117**, **118**, 137, 164
  composition 82
  copulation, role in 116, 164
  innervation 9, 116, 137
  muscles 136
  sense organ, role as 9, 116, 164
Spinneret 10
Starvation, body size 88
  metabolic rate 88
  oxygen consumption 88
Stephanurus, extra-corporeal digestion 52
Stichocytes 34, 48
Stomodaeum 28–36, **30**, **32**
Strongyloides, cuticle 146
  free-living generations 63–4, 65, **69**
  fumarate reductase pathway 69–70
  hormones, effect on 22
  lipid content 77
  parasite generations 63–4, 69–70, **69**
  PK/PEPCK ratio 69, **69**
  TCA cycle 65
Strongylus, enzymes 52
  haemoglobin 91
Stylet **frontispiece**, 3, 4, 15, 30, 31, 39–44, **40**, 47–8, 82, 126, **126**, 166–7
  animal parasitic 47–8
  carnivores 30, **30**, 31, 43–4
  composition 82
  entomophagous 45
  exploration by 41, 177
  feeding, role in 30, 31, 39–44, **40**, 166–7
  gustatory role 166–7
  hatching, role in 126, **126**
  moulting 15, 31
  muscles 136
  mycophagous 39–42, **40**
  odonto- 31
  penetration, role in **frontispiece**, **40**, 41, 42, 43–4
  plant parasitic **frontispiece**, **30**, 39–43
  stomato- 31
Sub-ventral glands (excretory system) **4**, **6**, 8, 9, 47, 52, 113–5, 139
  acetylcholinesterase in 114, 139
  enzymes in 47, 52, 114, 139
Swarming 173
Swimming 147–9, **149**, 151, **151**, **152**
Synapses 133, **134**, 138–40
  excitatory 140
  inhibitory 140
Synaptic transmission 138–42

Syngamus, anthelmintics, effect on 95
anticoagulant 56, 165
feeding **46**
haemoglobin 92–3, **92**
mouth parts 28–9, **30**
oxygen affinity **92**
Syphacia, anthelmintics, effect on 95
fumarate reductase pathway 69

Tanning 14–15, 82, 124–5
Taxis 168, **168,** 170–1
Teeth 3, 29, 30, **30,** 31, 44, 46, 47
Temperature, hatching, effect on 128–9, 132
metabolic rate, effect on 88
response to 88, 172–3
Testis **2,** 9, 116–7, **116**
Tetrameres, growth 26
Thelastoma, glycogen consumption 58
lipid consumption 77–8
Toxocara, ascarosides 79
hatching 90
Transfer cells *see* giant cells
Trehalase 51
Tricarboxylic acid cycle 60, 64–5, **66**
anthelmintics, effect on, 95–6
Trichinella, amino-acid uptake 84
cytochromes 72
dispersal 160
egg shell 124
electron transport 72
fumarate reductase pathway 69
glycogen content 58, **59**
lipid consumption 80
lipid content **77**
metabolism 58, 69, **69,** 72, 80, 111, 112
moulting 21
nitrogenous excretion **111,** 112
pharyngeal muscle 32
PK/PEPCK ratio **69**
sex attraction 119
stichocytes 34
Trichodorus, feeding 42
movement 154–5
muscles 134
roots, attraction to 176–7, **176**
Trichonema, light, response to 177
touch, response to 177
Trichostrongylus, acetylcholinesterase 114, 139
exsheathment 22–4
hatching 125, 128
movement 155
Trichuris, anthelmintics, effect on 95
bacillary band cells 8, 10
cytochromes 73
egg **123**
enzymes 51, 52
feeding 31, 47–8
fumarate reductase pathway 69, 73
stichocytes 34
stylet 31, 47
Turbatrix, electron transport 72
egg shell 124
feeding 39
fumarate reductase pathway 70
gluconeogenesis 75
glycogen consumption 59
glycogen content **59**
glyoxalate cycle 75, 80
growth *in vitro* 55
lipid consumption 59
lipid content **77**
metabolism 59, 65, 70, 72, 75, 80
moulting 21
swimming **148, 149**
TCA cycle 65
Tylenchorhynchus, feeding **40,** 42
Tylenchulus, dispersal 159
excretory system 125
feeding **40,** 43
gelatinous egg mass 114, 125
lipid consumption 77–8
protein consumption 82
Tylenchus, buccal cavity **30,** 31
cryptobiosis 151
feeding 44
see also phytophagous

Uncinaria, feeding 46
Undulatory propulsion 147–56, **148, 149, 151, 152, 157**
Urea, excretion 110, **111,** 113
Uterus **2, 3,** 9, 121, **121,** 125

Vagina **2,** 120–1, **121,** 125
Vas deferens **2,** 9, 38, **116,** 118
Volume regulation 97–102, **98, 100,** 101, 104–5
animal-parasitic nematodes 101–2, **101**
fresh-water nematodes 99–100
marine nematodes 97–9
plant-parasitic nematodes 100–1, **100**
soil nematodes 99–100

Water, hatching, effect on 127, 128
locomotion, effect on 147–55, **154**
permeability 103
transport **54,** 108
Wuchereria, diurnal rhythm 160
feeding 48

Xiphinema, feeding **frontispiece,** 35, 39
moulting 16
movement 153